T0251622

Assessing the Impacts of Environmental Changes on the Water Resources of the Upper Mara, Lake Victoria Basin

Assessing the Impacts of Environmental Changes on the Water Resources of the Upper Mara, Lake Victoria Basin

Promoters: Prof. Willy Bauwens
Prof. Piet N. L. Lens
Prof. Ann Van Griensven
Prof. Joy Obando

January 2014

Thesis submitted in fulfilment of the requirements for the award of the degree of
Doctor in de ingenieurswetenschappen (Doctor in Engineering) by

Fidelis Ndambuki Kilonzo

Promotors:
Prof.dr.ir. Willy Bauwens   Vrije Universiteit Brussel
Prof.dr.ir. Piet Lens   UNESCO-IHE & Wageningen University

Co-promotors:
Prof.dr. Joy Obando   Kenyatta University
Prof.dr.ir. Ann Van Griensven   Vrije Universiteit Brussel & UNESCO-IHE

PhD jury members:
Prof.dr.ir. Danny Van Hemelrijck, Chairman   Vrije Universiteit Brussel
Prof.dr ir. Stefan Uhlenbrook, Chairman   UNESCO-IHE & TU Delft
Prof.dr.ir. Marijke Huysmans, Secretary   Vrije Universiteit Brussel
Prof.dr. Joy Obando   Kenyatta University
Prof.dr.ir. Ann van Griensven   Vrije Universiteit Brussel & UNESCO-IHE
Dr.ir. Chris Mannaerts   ICT, Universiteit Twente
Prof.dr. Michael McClain   UNESCO-IHE & TUDelft

CRC Press/Balkema is an imprint of the Taylor & Francis Group, an informa business

© 2014, Fidelis Ndambuki Kilonzo

All rights reserved. No part of this publication or the information contained herein may be reproduced, stored in a retrieval system, or transmitted in any form or by any means, electronic, mechanical, by photocopying, recording or otherwise, without written prior permission from the publishers.

Although all care is taken to ensure the integrity and quality of this publication and information herein, no responsibility is assumed by the publishers or the author for any damage to property or persons as a result of the operation or use of this publication and or the information contained herein.

Published by:
CRC Press/Balkema
PO Box 11320, 2301 EH Leiden, The Netherlands
e-mail: Pub.NL@taylorandfrancis.com
www.crcpress.com – www.taylorandfrancis.com

ISBN: 978-1-138-02638-4

Acknowledgement

For a Journey that started way back in the dusty classes of Masewani Primary, Machakos County, Kenya, the PhD Journey has been long and torturous. It is a journey which I undertook and could not have been accomplished without the invisible Hand of God and the direct and indirect assistance of numerous persons.

With humility, I wish to thank my Promoters: Prof. Willy Bauwens (Vrije Universiteit Brussel, Belgium), Prof. Piet Lens (UNESCO-IHE, Netherlands), Prof. Ann Van Griensven (Vrije Universiteit/ UNESCO-IHE), and Prof. Joy Obando (Kenyatta University, Kenya) for the guidance, and inspiration and above all for the patience with my work and with me. I would like to thank my PhD defence committee chaired by Prof. Danny Van Hemelrijck (Vrije Universiteit Brussel, BE), and the Jury Members, Prof. Michael McClain (UNESCO-IHE, NL), Prof. Chris Mannaerts (Twente University, NL), Prof. Rik Pintelon (Vrije Universiteit Brussel, BE), and Prof. Marijke Huysmans (Vrije Universiteit Brussel, BE) for their critique and kind suggestions.

I appreciate the times shared with all members of the Hydrology and Hydraulics Engineering dept, VUB. Special mention of present and past PhD and Post-doctoral students including; Hepelwa, Olkeba, Tadesse, Narayan, Boud, Christian, Marwa, Jiri, Andualem, Khodayar, Zayir, Alvaro and Vicente. The support and services of; Anja, Hilde and Edward at the HYDR secretariat and that of J. Couder, Martine, Patrick and Carine at the VUB International office. I am also grateful for the periods I was at UNESCO-IHE, the Netherlands and the good time with PhD students; Frank, Njenga, Mukolwe, Getnet, Nagendra, Girma, Kiptala and Ezra, and for the dedicated assistance of Jolanda at the UNESCO-IHE fellowship office.

The PhD was made possible with the assistance of many players, individuals and organizations. I am ever grateful to the Flemish Inter-University council (VLIR, BE) and the National fellowship program (NUFFIC, NL) for granting me PhD fellowships. I am indebted to the administration of Kenyatta University for allowing for study leave, and the assistance from the Kenya Meteorological Dept, WRMA, DAO Bomet and Narok and line ministry officials. A big thank you to present and former members of staff in the Water & Engineering dept, Kenyatta Univesity; Dr S. Maingi, Dr C. Obiero, Dr W. Muthumbi, C. Githuku, P. Kimani, M. Wonge and F. Njagi.

Special appreciations to my family, my spouse and our children for bearing the burden and pain of my absence during my field work and the long hours spent on the PhD study.

Finally, in carrying out this study, materials, words and thoughts were obtained from different sources. It is not possible to detail in any precise way what each has done, but I sincerely hope that those concerned will accept this general acknowledgement in good faith. I am forever and sincerely grateful to you all and for everything.

Fidelis Ndambuki Kilonzo
Delft, January 2014

Abstract

Growing population and unregulated access to forest land have exerted high pressure on the land and water resources of the Upper Mara basin, leading to changes in land and water use patterns in the basin. This study considers the interactions among climate change and variability, land surface, hydrology, and human systems, including societal adaptations to changing environmental conditions.

The Upper Mara River catchment forms the recharge area for the Mara River basin, a key transboundary river, and one of the permanent rivers feeding into Lake Victoria. The area is drained by two main rivers: the Amala and the Nyangores which merge at the middle reaches (1°2'15.2"S, 35°14'31.7"E) to form the Mara River. The study aims to assess how changes in climate, land use and management practices have impacted on the water resources of the Upper Mara basin.

The objectives of the study are: to assess the trends in changes in the climatic, land cover/land use, water quality and vegetation variables; to build and evaluate a hydrological model capable of simulating the response of watershed processes to changing climatic, land use, and management conditions under past, present conditions; and predict potential impacts of the changes in climate, land use and management practices, and contribute in advising policy in the formulation and development of strategies aimed at the sustainable management of water resources in the Mara River Basin.

Historical data including data for rainfall, temperature and streamflow; field collected data; and satellite remote sensing data is used in the study. The Soil and Water Assessment Tool (SWAT) is used to evaluate the impacts of the changes in climatic, landcover and management inputs. Changes are made to both the model crop database and management files to make it adaptable to tropical conditions. The model is calibrated using streamflow data, and validated using both streamflow and distributed data. The performance of the model is statistically assessed to be "good" and "satisfactory" for the percent bias and Nash Sutcliffe respectively. The water balance fractions are within typical hydrological ranges and significantly similar to the observed fractions. Distributed validation using remote sensed leaf area index (RS_LAI), shows that the timing of the start of the growing season match well with SWAT simulated LAI.

Trend analysis is performed using the Mann-Kendall and Sen Statistics for rainfall and temperature variables. Rainfall data (1962-2009) from six stations located within the basin while temperature data (1992-2009) from three meteorological stations surrounding the study area is used. There is a general decreasing trend of upto 18mm/yr in the annual rainfall. 50% of the stations analyzed experienced significant decline (at $\alpha = 0.1$) in annual Rainfall in the last 50 yrs. While the minimum temperatures have significantly increased by between 0.02 and 0.04 ^{0}C/yr; there has been an upward but insignificant change in the maximum temperatures. Changes in the vegetation indices are analyzed using the normalized difference vegetation index (NDVI) data (1998-2010) from the Satellite Pour l'Observation de la Terre- VEGETATION (SPOT_VGT) sensor. All indices including, the integral NDVI, the vegetation condition index (VCI), the standardized vegetation index (SVI) and the vegetation productivity index (VPI) have a better than average vegetation health for different vegetation types showing an increasing trend in vegetation biomass.

Land cover change analysis using post classification methods, show that between 1976 and 2006, agriculture coverage increased by 109%, while forest, shrubs, and grassland decreased by 31%, 34%, and 4% respectively. Land change evolution using the three date (1986,1995,2006) NDVI-RGB method show that the highest loss in vegetation biomass was experienced in the 1990s. The biggest loser was the encroachment and degradation of the Mau forest areas.

Future management scenarios simulating the application of fertilizer at a typical rate of 100kg/Ha lead to significant increase in crop yields for all but two of the soil classes. The two soil classes have no marked improvement in yields even with high fertilizer application. The re-afforestation scenario to pre-1976 forest coverage results in reduction of streamflow and increase in evapotranspiration. In the irrigation scenario crop yields increase two to three fold even without fertilizer application. The replacement of the lowland cereal crops with grain sorghum led to higher water yields in the stream. Sorghum yield were comparable to the maize yields and the water stress days were halved in the sorghum scenario.

Sensitivity analysis of the climatic variables with SWAT indicate that on one hand, higher watershed evapotranspiration (ET) will result from higher rainfall and temperatures, with higher CO_2 leading to decline in ET. On the other hand higher rainfall and higher CO_2 will also lead to high water yields, but higher temperature will lead to a drop in the water yield.

The Long Ashton Research Station Weather Generator (LARS-WG) is used in the weather perturbations, and the construction of the GCM climate change scenarios. The projection of climate change with ensemble mean for 16 GCMs in the 2020s and 2050s show increase precipitation amounts compared to 1990s baseline. The 2050s will have higher projected rainfall than 2020s, while higher rainfall is projected under B1 than A1B IPCC climate scenarios. As a consequence, higher streamflow is expected for all future time periods. The return level at the outlet of the basin for a presumed 30, 50 and 100 yr flood was determined to increase by 11% in the 2020s to 19% in the 2050s. Bias corrected GCM outputs however indicate marginal increases in the different return levels (3% for 100yr flood) but more severe extreme low flows.

Whereas the historical trends depict a tendency towards a drier watershed, the uncorrected climate models predict a wetter future. This inconsistencies present a challenge in planning for future management of water resources at the Mara basin. Bias corrected climate projections indicate a trend which was more constistent with historical records. As an adaptation measure, the conservation of the flood water even at current flow levels can go a long way in alleviating water stress in the dry months. In the short term the adoption of better farm management practices including appropriate use of fertilizers and introduction of drought tolerant grain crops like sorghum should be encouraged.

Table of Contents

List of figures

List of tables

List of abbreviations and acronyms

AAFC-WG	Agriculture and Agri-Food Canada weather generator
AGNPS	Agricultural Non-point Source
AnnAGNPS	Annualized Agricultural Non-point Source
ANSWERS	Area Non-point Source Watershed Environment Response Simulation
AR4	Fourth Assessment Report
ASAL	Arid and semi-arid lands
ASARECA	Association for Strengthening Agricultural Research in Eastern and Central Africa
AVHRR	Advanced Very High Resolution Radiometer
BSAP	Biodiversity Strategy and Action Plan
CA	Cluster and average method
CAN	Calcium ammonium nitrate
CFR	Central Forest Reserves
CIMMYT	Centro Internacional de Mejoramiento de Maíz y Trigo (International Maize and Wheat Improvement Center)
CYCLOPES	Change in Land Observational Products from an Ensemble of Satellites
DAO	District Agricultural Officer
DAP	Di-ammonium phosphate
EAC	East African Community
EFA	Environment Flow Assessment
ENSO	El Niño/Southern Oscillations
EPIC	Erosion Productivity Impact Calculator
ETM+	Enhanced Thematic Mapper Plus
FAO	Food and Agriculture Organisation
FD	Forest Department
GCM	General Circulation Models
GLUE	Generalized Likelihood Uncertainty Estimation method
GOK	Government of Kenya
GSSHA	Gridded Surface Hydrologic Analysis
HEC-HMS	Hydrologic Engineering Center's Hydrologic Modeling System
HRU	Hydrologic response unit
HSPF	Hydrological Simulation Program – FORTRAN
HydroSHEDS	Hydrological data and maps based on SHuttle Elevation Derivatives at multiple Scales
HYMO	Hydrologic model
ICPAC	IGAD Climate Prediction and Applications Centre
IFA	International Fertilizer Industry Association

IFDC	International Fertilizer Development Center
IGAD	Intergovernmental Authority on Development
iNDVI	Integrated NDVI
IPCC	Intergovernmental Panel on Climate Change
ISODATA	Iterative Self-Organizing Data Analysis Technique
ITCZ	InterTropical Convergence Zone
IWRM	Integrated Water Resources Management
KENSOTER	Kenya Soil and Terrain
KINEROS2	KINematic Runoff and EROSion
KNBS	Kenya National Bureau of Statistics
LAI	Leaf area index
LARS-WG	Long Ashton Research Station-weather generator
LCCS	Land Cover Classification System
LH-OAT	Latin Hypercube – One At a Time
LVBC	Lake Victoria Basin Commission
LVBWO	Lake Victoria Basin Water Office
MAM	March-April-May
MCMC	Markov chain Monte Carlo
MLN	Maize leaf necrosis disease
MMD	Multi-model dataset
MODIS	Moderate Resolution Imaging Spectroradiometer
MoU	Memorandum of understanding
MRCC	Maritime Rescue Coordination Centre
MSS	Multi spectral scanner
(M)USLE	(Modified) Universal Soil Loss Equation
NDVI	Normalized Difference Vegetation Index
NRC	Non-Resident Cultivation
NSE	Nash-Sutcliffe efficiency
OLS	Ordinary Least Squares
OND	October-November-December
ParaSol	Parameter Solution
PBIAS	percent bias
PCMDI	Program for Climate Model Diagnosis and Intercomparison
PES	Payment for Ecosystem Service
PET	Potential evapotranspiration
PRMS	Precipitation-Runoff Modeling System
PSO	Particle Swarm Optimization

RREL	Relative range of the NDVI
RSR	Root Mean Square Error observations standard deviation ratio
SEA	Strategic Environment Assessment
SHE	Systéme Hydrologique Européen
SPOT-VGT	Satellite Pour l'Observation de la Terre-VEGETATION
SRTM	Shuttle Radar Topography Mission
STHP	Sub Tropical High Pressure systems
SUFI2	Sequential Uncertainty Fitting-version 2
SVI	Standard Vegetation Index
SWAT	Soil and Water Assessment Tool
SWAT-CUP	SWAT - Calibration and Uncertainty Programs
TM	Thematic Mapper
TN	Total nitrogen
TP	Total phosphorus
TWRUF	Trans-boundary Water Resources Users Forum
UNEP	United Nations Environment programme
UNFCCC	United Nations Framework Convention on Climate Change
VCI	Vegetation Condition Index
VGT4Africa	Vegetation for Africa project
VITO	Vlaamse Instelling Voor Technologisch Onderzoek (Flemish Institute for Technological Research)
VPI	Vegetation productivity indicator
WEPP	Water Erosion Prediction Project
WGN	Weather generator
WHO	World Health Organization
WMO	World meteorological Organization
WRMA	Water Resources Management Authority
WRUA	Water Resources Users Association
WUA	Water Users Association
WWF-ESARPO	World Wildlife Fund-East and Southern Africa Programme Office
WXGEN	Weather generator

1. General introduction

1.1 Introduction

Multiple stresses and the low adaptive capacity has led to high vulnerability of the African content to changes and variability in climate. (Boko et al., 2007). Over reliance of rainfed agriculture compounded by the extreme levels of poverty and disasters have all contributed to the scenario of Africa's high vulnerability. Due to climate change, agricultural losses are shown to be possibly severe for several areas, where yields from rain-fed agriculture could be reduced by up to 50% by the year 2020. Some assessments by the Intergovernmental Panel on Climate Change (IPCC) project that, by 2020 between 75 and 250 million people will be exposed to increased water stress due to climate change (Parry et al., 2007).

According to the IPCC, the arid and semi-arid areas are most susceptible to changes in climate which could worsen the water scarcity problem in these areas. Climate change influences the timing and magnitude of runoff and sediment yield. Changes in variability of flows and pollutant loading that are induced by climate change have important implications on water supplies, water quality, and aquatic ecosystems of a Watershed (Prathumratana et al., 2008). Lack of access to safe water, arising from multiple factors, is a key vulnerability in many parts of Africa.

In Kenya, "Climate models predict an increase in climate variability, indicating that Kenya's vulnerability is set to get worse. Agriculture, tourism, health, energy, transport and infrastructure, water supply and sanitation are the sectors expected to be those most severely affected by climate change in the long term" (Downing et al., 2008). Growing population exerts high pressure on the limited land and water resources in the Mara basin, leading to changes in land and water use patterns in the basin. Vörösmarty, et al., 2000 noted that ,"to secure a more complete picture of future water vulnerabilities, it will be necessary to consider interactions among climate change and variability, land surface and groundwater hydrology, water engineering, and human systems, including societal adaptations to water scarcity".

According to Urama et al., 2008, "the current lack of any explicit policy instruments that take a holistic approach for the whole of the Mara River ecosystem highlights the urgent need for the integration of the social, economic and ecological aspects of water management in a single,

1

trans-boundary resource management policy". Urama and Davidson (2008) noted that "there is still much work ahead for a sustainable integrated trans-boundary river management policy to be developed for the Mara River".

Several strategic documents have been adopted by the Lake Victoria Basin Commission (LVBC) of the East African Community (EAC) within its activities to promote regional coordination in trans-boundary water resources management (Onyando et al., 2013). These include: the Environment Flow Assessment (EFA) for the Mara River (LVBC & WWF-ESARPO, 2010a), the Biodiversity Strategy and Action Plan (BSAP, LVBC & WWF-ESARPO, 2010b), the Strategic Environment Assessment (SEA) for the Mara River Basin (LVBC et al., 2012), and Payment for Ecosystem Service (PES, Bhat et al., 2009) policy guide for the transboundary Mara River Basin.

The Trans-boundary Water Resources Users Forum (TWRUF) was created in an attempt to promote regional policies on trans-boundary natural resources management within the EAC (Onyando et al., 2013). Successful integrated water resources management (IWRM) implementation requires good water governance supported by institutions that can administer WRM effectively at all levels. The water sector reforms in Kenya and Tanzania stipulate the formation of water institutions at all levels. Examples include Water Resources Management Authority (WRMA) and Water resources Users Associations (WRUAs) in Kenya, the Lake Victoria Basin Water Office (LVBWO) and Water Users Association (WUAs) in Tanzania. The water sector reforms together with successful lobbying of key stakeholders to endorse grassroot water resource management institutions (WRUAs/WUAs) through memoranda of understanding (MoUs) are key steps towards building their legal, technical and financial sustainability in water governance (Onyando et al., 2013).

Despite these developments, there are knowledge and institutional gaps that still need to be addressed specific to the Upper Mara and in general for the entire Mara basin. For example, literature searches reveal that there is no coordinated water quality monitoring scheme for the Mara River Basin. Gann (2006), while addressing the data scarcity problem in the study area characterized data as "non-existence, non accessible, limited accessible, with gaps, not very accurate, and in different formats and no standards". Extensive efforts have been made in this study to improve on the usability of the available data sets for modeling purposes and explore the use of alternative data sources for both trend analysis and model validation.

1.2 Justification and significance of the research

In the absence of good data (quantity and quality), conventional methods of studying an area using only standard observed hydro-meteorological data may not yield the desired results. This study therefore explores the use of modeling tools, field studies and satellite remote sensed data to study the hydrological processes of the Upper Mara basin. The findings of this research will make a contribution towards the sustainable management of the Mara river basin by providing additional information, insights and perspective. The Kenya strategic development blueprint "Vision 2030", identifies several flagship projects in the Water and Sanitation sector (GOK, 2007). Amongst these projects the rehabilitation the hydro-metrological network, the construction of multi-purpose dams with storage capacity of 2.4 billion m^3 along rivers Nzoia and Nyando, and the construction of medium-sized multi-purpose dams with a total capacity of 2 billion m^3 to supply water for domestic, livestock and irrigation use in the arid and semi-arid lands (ASAL) areas have a direct bearing on the Upper Mara region and the Lake Victoria basin as a whole.

Planned research questions will seek to investigate the impacts of some of these projects on the water resources and economic productivity in the study area. In addition to the scientific output the study will strive to make its contributions towards the realization of the development of a holistic management strategy and policy for the Mara River basin.

1.3 Objectives and research questions

The study aims specifically to:

1. Assess the trends in the changes in climate, land cover/land use and vegetation variables in the Mara River Basin

2. Adapt a process based hydrological model to evaluate and predict the response of hydrological processes to changing climatic, land use and water management conditions under past, present and future conditions

3. Assess the impacts of climate, land use and water management changes on the sustainable management of water resources in the Mara River basin.

3

In order to meet the set objectives, several research questions were formulated to guide in attaining the expected outputs of the PhD study. The study is therefore based on the following research questions:

1. Has there been a significant variation in monthly, seasonal and annual precipitation amounts and frequency in the last 50 years?
2. Have the minimum and maximum temperatures changed significantly over the last 20 years?.
3. Has the health of the vegetation significantly changed over 12 year period?
4. Is the SWAT hydrological model adequately suited to simulate hydrological processes in tropical African catchments like the Upper Mara basin?.
5. Has the conversion of land from one land cover and use to another affected the local water quality, hydrological and climate regimes?.
6. How are projected changes in climate likely to affect the hydrological regime of the Upper Mara river basin?.
7. What are the potential adaptation strategies towards the established change, for both the preservation of the ecosystem and for the improvement of livelihoods?.
8. Which best management practices are effective for the Mara basin considering the projected global changes?.

The PhD is structured into five building blocks:

- the introduction part;
- the data analysis part;
- the hydrological modeling part;
- the scenarios management part, and
- Conclusions

While each block addresses a specific objective, the components build stepwise on each other to produce a coherently flowing storyline, conclusive at each stage and hence minimizing redundancy and repetitiveness.

1.4 Overview of the structure of the thesis

There are twelve chapters in the five building blocks. The first block consists of an introduction and a review of the relevant literature pertaining to climate and land use change as well as a review of hydrological models for application in tropical climates. Also, previous applications of the Soil and Water Assessment Tool (SWAT) simulator – the hydrologic simulator that will be used in our study- in the lake region and the larger Nile basin are reviewed.

The second part describes the study area and the historical trends that have characterized the physical, social and economic environments of the study area. The data sections look at the available quality and quantity of data and explore the use of alternative sources of data to compensate for the missing data. Next, the interpretative, processing and manipulation procedures that were applied in order to make the data usable for the various applications in this study are discussed.

The hydrological modeling part involves the set-up of the SWAT model, the adaptation of the model to suit tropical African catchments and the calibration and validation of the model.

Once the model was deemed satisfactory and adequate for application in the watershed, future climate, water management and land cover change scenarios were used on the model to assess their impact on the river flow and the hydrology of the Upper Mara river basin. The results of the modeling of future scenarios are used to discuss potential strategies which might inform policy on the Mara river water resources management.

The concluding chapter looks at the key findings from the study, the shortcomings and challenges experienced in trying to achieve the objectives. The limitations of both the adapted model and the study findings are also discusssed. The study offers some recommendations going forward for better modeling results in similar catchments and data situation.

2. Climate and land use change studies

2.1 Climate change

There are differing definitions of the term climate change. On the one hand, the IPCC, 2001 defines climate as "a statistically significant variation in either the mean state of the climate or in its variability, persisting for an extended period, typically decades or longer". Climate change may be due to natural internal processes or external forcings, or to persistent anthropogenic changes in the composition of the atmosphere or in land use. The United Nations Framework Convention on Climate Change (UNFCCC), on the other hand, defines climate change as "a change of climate which is attributed directly or indirectly to human activity that alters the composition of the global atmosphere and which is in addition to natural climate variability observed over comparable time periods". The UNFCCC, therefore, uses the term Climate Change to indicate only those changes brought about by anthropogenic causes.

There is a growing consensus among scientific and political leaders that climate change is the biggest environmental threat modern society faces. According to the scientific opinion, there has been a sustained increase in global average temperatures that began to have an effect on the earth's climate. According to the IPCC, 2001, the average temperature of the earth's surface has risen by 0.6°C since the late 1800s. It is expected to increase by another 1.4 to 5.8°C by the year 2100. Some investigators have come to the conclusion that even if the minimum predicted increase takes place, it will be larger than any century-long trend in the last 10,000 years (IPCC, 2001). The growing concentration of greenhouse gases causes a gradual rise in temperature, and for many areas in the world, impacts on precipitation (rain and snow) patterns, and on the frequency of extreme events, such as extreme temperature, rain storms, droughts, and consequently also on the risk of flooding and low flow effects (IPCC, 2001). The average sea level rose by 10 to 20 cm during the 20th century (Wood et al., 2004), and an additional increase of 9 to 88 cm is expected by the year 2100 (IPCC, 2007).

According to IPCC (2007), the fourth assessment report (AR4) found that "warming of the climate system is unequivocal based on an increasing body of evidence showing discernible physically consistent changes. These include: increases in global average surface air temperature; atmospheric temperatures above the surface, surface and sub-surface ocean water temperature; widespread melting of snow; decreases in Arctic sea-ice extent and thickness; decreases in glacier and small ice cap extent and mass; and rising global mean sea level. The observed surface warming at global and continental scales is also consistent with reduced duration of freeze seasons; increased heat waves; increased atmospheric water vapor content and heavier precipitation events; changes in patterns of precipitation; increased drought; increases in intensity of hurricane activity, and changes in atmospheric winds". These changes will translate into changes in the hydrological cycle which will show different regimes totally depending on the climate behaviour.

The most comprehensive way to infer future climatic change associated with the perturbation of atmospheric composition is by means of three-dimensional General Circulation Models (GCM) (Cess et al., 1990). The GCMs have been developed to simulate the present climate and have been used to predict future climatic change (Xu, 1999). The GCM are restricted in their usefulness for many subgrid scale applications by their coarse spatial and temporal resolution (Wilby and Wigley, 1997; Lettenmaier et al., 1999, Wood et al., 2002, and Wilby et al., 2000). While GCMs have a resolution of 150-300 km by 150-300 km, many impact models require information at scales of 50 km or less. Some method is therefore needed to estimate the smaller-scale information.

Errors in the parameterizations of both GCMs and hydrological models affecting both the temporal and spatial dimensions occur on the scale(s) at which climate and terrestrial impact models interface. These errors are a key source of uncertainties and have important implications for the credence of impact studies derived by the output of models of climate change, especially due to the fact that research into potential human-induced modifications to hydrological and ecological cycles is assuming increasing significance (Montanari, 2007). To address this challenges, tools for generating the high-resolution meteorological inputs required for modeling ecohydrological processes are needed.

Downscaling approaches can be used to relate large-scale atmospheric predictor variables to local or station-scale meteorological series. "Downscaling is the process of deriving finer resolution information from larger scale weather and climate model output for use in hydrologic modeling and water resources management applications" (Georgakakos et al., 2006). Downscaling methods are classified as either statistical or dynamical. There are three categories of Statistical downscaling (SD) techniques: transfer functions, weather typings and weather generators (Chen et al., 2011). In the transfer function approach, statistical linear or nonlinear relationships between observed local climatic variables (predictands) and large-scale GCM outputs (predictors) are established. They are relatively easy to apply, but their main drawback is the probable lack of a stable relationship between predictors and predictands.

Weather typing schemes involve grouping local meteorological variables in relation to different classes of atmospheric circulation. The main advantage is that local variables are closely linked to global circulation. However, its reliability depends on a stationary relationship between large-scale circulation and local climate. Especially for precipitation, there is frequently no strong correlation between daily precipitation and large-scale circulation. The weather generator (WG) method is based on the perturbation of it's parameters according to the changes projected by climate models. The appealing property is its ability to rapidly produce sets of climate scenarios for studying the impacts of rare climate events and investigating natural variability.

Dynamical downscaling methods involve dynamical models of the atmosphere nested within the grids of the large scale forecast models. One way nested limited area weather or regional climate models are implemented to produce finer resolution gridded information for applications, with coarse resolution models providing initial and lateral boundary conditions. Murphy (1998), in his study of downscaling estimates for 976 European stations from June 1983 to February 1994, concluded that the dynamical and statistical methods perform with similar skill in downscaling observed monthly mean anomalies. According to Xu, 1999, recent higher-resolution regional climate models provide better agreement with observations on synoptic and regional scales and on monthly, seasonal and inter annual timescales. Examples include statistical downscaling approaches that link GCMs to meteorological and hydrologic models resolved at finer scales.

2.2 Land use and Land cover changes

Although the terms "Land use" and "Land cover" are often used interchangeably, each term has a very specific meaning with some fundamental differences. Landcover on the one hand denotes the biophysical cover over the surface including such features as vegetation, urban infrastructure, water, bare soil or other. It does not describe the use of land, which may be different for lands with the same cover type. On the other hand, land use refers to the purpose the land serves, and describes human influence of the land, or immediate actions modifying or converting land cover (De Sherbinin, 2002, Ellis, 2009). Land cover therefore, is the "physical state of the earth's surface and immediate subsurface, while land use involves both the manner in which the biophysical attributes of the land are manipulated and the intent underlying that manipulation" (Turner et al., 1995).

Distinction may also be made between changes in land cover and changes in land use. Changes in land cover leads to change in cover type (forest to pasture, cropland to woodland, agriculture to urban), and change in cover characteristics (structure, field size, degradation, productivity). Changes in land use mean change in land management practices or ownership, intensification, mechanization, irrigation, abandonment, cropping system. According to Ellis (2010), land-use and land-cover change (LULCC) is a general term for the human modification of the earth's terrestrial surface. LULCC modifies surface albedo and thus surface-atmospheric energy exchanges, which have an impact on the regional climate. Since terrestrial ecosystems are sources and sinks of carbon, any change in land-use/cover impact the global climate via the carbon cycle. The contribution of local evapotranspiration to the water cycle (precipitation recycling) as a function of landcover also impacts the climate at a local to regional scale (Lambin et al., 2003). According to Lambin et al. (2003), LULCC is driven by a combination of fundamental high-level causes: a) resource scarcity leading to an increase in the pressure of production on resources, b) changing opportunities created by markets, c) outside policy intervention, d) loss of adaptive capacity and increased vulnerability, and e) changes in social organization, in resource access, and in attitudes.

Changes in the Land-use and land-cover have serious and far reaching consequences including altering the earth system functioning. According to Lambin et al., 2001, these changes directly impact biotic diversity worldwide, contribute to local and regional climate change as well as to global climate warming, are the primary source of soil degradation, and by altering ecosystem services affect the ability of biological systems to support human needs. Estimates of the areal extent, spatial expression or likewise quantitative estimate of the impact of land change more or less converge, while estimates driven by notions are larger and dramatic

(Lambin and Geist, 2006). In order to assist in analysis and understanding of land use dynamics, several types of Land use change models have been developed. Land use change does not occur evenly, neither temporarily nor spatial. Estimates over the last 300 years approximate losses of 10-30% of forests and woodlands, 1% of grasslands and grasslands pastures, while cropland areas increased by 466% (Richard, 1990, Kees, 2001).

The best area-efficient method for studying land cover and land cover changes is remote sensing. The combination of remote sensing information with other available enabling technologies including the global positioning system (GPS) and the geographic information system (GIS), maybe used as relatively inexpensive tools to assist in planning and decision making (Franklin et al., 2001, Rogan and Chen, 2004). Advances in remote sensing data acquisition and interpretation techniques has contributed to the knowledge base on land cover changes. Freely available global earth observation products have provided ways to achieve rapid assessment and monitoring of land change hotspots at the landscape scale.

Lepers et al. (2005) synthesized 49 data sets available in early 2003 at the national and global scale to identify locations of rapid land-cover change. Some of these data sets identified hotspots of land-cover change, and others provided estimates of rates of change. For the estimates of rates of change areas, the highest change rates were identified by applying a threshold percentile value. Threshold values were determined for each of these data sets to identify the areas having a high percentile in terms of rates of change. The study established that deforestation is the most measured process of land-cover change at a regional scale. During the 1990s, forest-cover changes were much more frequent in the tropics than in the other parts of the world.

The main areas of recent cropland increase are spread across all continents. Some areas of decrease in cropland extent are located in the other continents, except for Africa, where no decrease in cropland area was identified (Lepers et al., 2005). The region of the great lakes of eastern Africa is one of the principal locations where cropland expansion has taken place (Leper et al., 2005). Maitima, 2009 noted that "Land use changes in East Africa have transformed land cover to farmlands, grazing lands, human settlements and urban centers at the expense of natural vegetation. These changes are associated with deforestation, biodiversity loss and land degradation".

The Mara River basin has also experienced considerable vegetation changes. For instance,according Seernel et al., 2001, the ecosystem in the south-eastern part of the Kenyan side of the basin (present day Narok south district) has passed through successive stages of transformation as the result of the interaction between four distinct, and probably cyclical, processes of change: i) change in vegetation, ii) change in climate, iii) tsetsefly and tick infection and iv) pastoral occupation and management. Land cover change analyses have been undertaken in the Mara River basin using diverse sources of satellite remote sense data and change analysis techniques by Serneels et al. (2001). Mati et al. (2008), and Mundia et al. (2009).

Table 2-1 summarizes the temporal coverage, the datasets and the key findings of these landcover change studies. In all these studies as is consistent with others in the region, the area under cropland has steadily increased at the expense of other land cover types, especially forest and grassland.

Table 2-1: Previous land change analysis by different authors in the Mara river basin.

Author(s)	Period	Dataset used	Change detection method.	Key findings
Seernels et al., 2001	1975 to 1995	AVHRR GAC MSS (29-07-1975) TM (09-01-1985) TM (21-01-1995)	Image differencing and change vector Magnitude CVM)	1. 3 times more vegetation loss in 1985-1995 than 1975-1985 2. Landcover change caused by expansion of large scale wheat cultivation
Mati et al., 2008	1973 to 2000	MSS 1973 and TM 1986, ETM 2000	Post classification	Agriculture increased by 50% Rangelands decreased by 27% Forest decreased by 32%
Mundia et al., 2009	1975 to 1986	MSS 11-02-1975, TM 17-10-1986 AVNIR/ALOS	Post classification	Agriculture expansion by 12% Forest decreased from 11 to 9%

2.3. Conclusion

There is overwhelming scientific evidence that the global climate is changing. The changes are however not even across the globe. The impacts of these changes in the global climate on the hydrology of a river catchment can only be made possible with downscaled data. The use of statistical downscaling tools including weather generators will provide the much needed climate projection data

Land use change has been greatly driven by the need to produce more for the ever increasing population and changing lifestyle needs. Recent development in social and environmental awareness in the developed nations has seen a reversal of this trend. However in the developing countries and especially those in Africa and Asia, natural vegetation is still being lost to agriculture. The global trend has been repeated in the Mara River basin with significant area under forest and grassland being converted to subsistence and commercial farming. Natural wildlife habitats have also been taken up by herding communities restricting the wildlife movement to the nature reserves only. There is a need to further explore on the changes taking place in the Mau water towers, the source of the Mara River basin which has been addressed only in limited previous studies.

3. SWAT hydrologic modeling in the larger Lake Victoria region

3.1 Introduction

The SWAT simulator (Arnold et al., 1998) has been used extensively in Africa for a variety of applications. Van Griensven et al. (2011) reviewed 36 journal papers touching on SWAT in Africa, 60% of these were from the Nile basin region, with very little work in West Africa region. The applications of SWAT on the African continent include: the model performance analysis (calibration, validation, uncertainty analysis and SWAT model development), land management impacts, erosion and sediment yield analysis, water quality analysis and the assessment of the impacts of climate change and land use / land cover change (LULCC) impacts on the hydrology and the water balance. Other crosscutting applications include the assessment of economic productivity tradeoffs. Ndomba and Birhanu (2008) reviewed the complexity and challenges of SWAT modelling in Eastern Africa and found that SWAT satisfactorily simulates river flows in study catchments with limited data availability and where global spatial data are appropriate.

3.2 Hydrologic modelling

Jayakrishnan et al. (2005) modelled the hydrology of the 3050km^2 Sondu river basin with limited data on land use, soil and elevation, with the aim of assessing the impacts of land use changes as a result of changes of intensive dairy farming. The model performance with a Nash-Sutcliffe efficiency (NSE) of <0.1 was attributed to inadequate rainfall and other model input data. They concluded that "use of one rain gauge station situated at the upper end of the catchment was not representative of the basin".

Mulungu and Munishi (2007) calibrated a SWAT model for the 11000 km^2 Simiyu catchment using improved spatial inputs for land use and soil. The study used a land use map developed from Landsat thematic mapper (TM) images. Local soil and geological maps were used to augment the SOTER 1:2000000 global databases. Although the resulting water yield and the surface runoff fraction of the water balance were within ±1% of the observed flow, the base flow fraction was off target by 50%. According to the authors, improvements of the spatial resolution of the soil and land use inputs did little to improve the model performance. The results indicate that the SWAT model maybe more sensitive to climatic inputs that other spatial inputs.

In modeling the hydrology of the Mitano river basin in Uganda, Kingston and Taylor (2010) used the gridded 0.5° CRUTS3.0 database as the climatic input. They found a good agreement between observed and simulated monthly means and flow duration curves, though the model performance after calibration was poor with an NSE of -0.09. The poor performance of the model was attributed to "model-observation divergences with the calibration period simply too large to be resolved by an autocalibration routine". The gridded 0.5° CRUTS3.0 database used as model input for the hydrological modeling was referred to by the authors as "of questionable accuracy over the Mitano basin". The use of global coarse resolution databases for the hydrological modeling of catchments in the region has also been questioned by other authors including Jacobs and Srinivasan (2005) for the Upper Tana basin in Kenya and Haguma (2007) for the Kagera basin.

3.3 Modeling of land use and land cover changes

Githui et al. (2009a) analysed the impacts of land cover change on the runoff for the Nzoia basin in Kenya. The emphasis was on "reforestation and sustainable agriculture" as a best-case scenario, and "deforestation and expansion of unsustainable agriculture" as a worst-case scenario. The worst case scenario yielded more runoff, baseflow and streamflow, while the best case yielded reduced amounts of the same variables. The increase in streamflow was attributed to decreased evapotranspiration and increased surface runoff. They concluded that the increase in runoff potent a higher likelihood of flood-like events.

Mango et al. (2011) used three hypothetical scenarios: partial deforestation, complete deforestation to grassland and complete deforestation to agriculture, to analyse the sensitivity of the model outputs to land use change for the Nyangores (700km^2), a tributary of the Mara river basin in Kenya. Simulations under all land use change scenarios indicated a reduced baseflow and average flow.

3.4 Modeling of climate change impacts

Kingston and Taylor (2010) explored the impacts of projected climate changes on the water resources of the Upper Nile basin and on the 2098 km^2 Matano basin in Uganda. The assessment included the evaluation of the range of uncertainty due to climate sensitivity, choice of General

Circulation Models (GCMs) and hydrological model parameterization. Results of the uncertainty analysis indicated that model parameterization generally imparts little uncertainty to the climate change projections compared to the GCM structure. The authors found an overwhelming dependence upon the GCM used for climate projections and showed that single-GCM evaluations of climate change impacts are likely to be completely inadequate and potentially misleading as a basis for the analysis of climate change impacts on freshwater resources. On the hydrology, the study found that the proportion of precipitation that contributes to the Mitano river discharge via groundwater will decrease as a result of increasing temperature. The increasing evapotranspiration due to increasing global temperatures limits the amount of water penetrating the soil profile and replenishing the shallow groundwater store during the wet season rather than reduced precipitation.

Githui et al. (2009b) used monthly change fields of rainfall and temperature instead of mean annual perturbations to the historical time series or hypothetical scenarios for the 12700 km^2 Nzoia basin in Kenya. Scenarios of future climate were obtained by adjusting the baseline observations by the difference for temperature or percentage change for rainfall between period-averaged results for the GCM experiments (30-year period) and the simulated baseline period (1981–2000). The A2 scenario gave more increases in rainfall than B2 in each time period. According to these scenarios, more rainfall will be experienced in the 2050s than in the 2020s, the seasonality of rainfall will still be maintained even though the total amounts vary. All the scenarios indicated that temperature would increase in this region, with the 2050s experiencing much higher increases than the 2020s. The models were consistent with respect to changes in both runoff and baseflow, with average streamflow observed to increase with rainfall increase, relatively higher amounts were observed in the 2050s than in 2020s. All scenarios indicated higher probabilities to exceed the bankfull discharge than the observed time series. The A2 scenario projected a higher number of flood-like events than B2.

Mango et al. (2011) developed regional temperature and precipitation projections from a set of 21 global models in the MMD for the A1B scenario for East Africa. The hydrological model was run for minimum, median and maximum change scenarios. Notable is the nonlinear response, with large stream flow changes occasioned by only small changes in precipitation. A combined decrease in precipitation and an increase in temperature led to increased evapotranspiration and reduced runoff.

Dessu and Melesse (2012) evaluated sixteen GCMs to assess the impact and uncertainty of climate change on the hydrology of the Mara River basin and further selected five of them for the assessment of future climate scenarios in the basin. On the basis of the observed seasonal variation, the raw GCMs outputs showed inferior skill to capture the bimodal tropical rainfall pattern. The average flow hydrograph of five selected GCMs showed an increase in the flow volume of the Mara River both in the 2050s and the 2080s. Compared with the control period, the hydrologic regime may experience a tremendous pressure due to extreme high and low flows where the wet seasons become wetter and the dry seasons might probably become drier.

Whereas Githui et al. (2009) argues that stream flow response was not sensitive to changes in temperature, Kingston and Taylor (2010) and Mango et al. (2011) argue that increases in temperature will lead to an increase in evaporation and hence a change in the water balance, reducing the stream flow. It should be noted that, both Kingston and Taylor (2010) and Mango et al., (2011) used satellite derived climatic data as their input into the hydrological model and baseline, while Githui et al. (2009) built their model on observed climatic data. The conflicting results imply high sensitivity of hydrological processes to climate variable characteristics and make a case for improving the quality of the observed rainfall data rather than using satellite rainfall. The gridded satelite rainfall data is based on global databases on a relatively coarse resolution. Another difference between the two approaches is the size of the catchments under consideration. Whereas all hydrological processes may not have been closed in the small sized catchments of the Mitano and Nyangores (<2100 km^2), the large sized catchment (Nzoia >12000 km^2) enables the completion and closing of all hydrological processes.

3.5 Cross-cutting applications

Swallow et al. (2009) used the SWAT model to estimate sediment yields for two basins draining into the Lake Victoria from the Mau region in Kenya. Each sub-basin was then identified as belonging to one of the four categories. The authors noted "the inability of the SWAT model to consider gully in the Modified Unified Soil Loss Equation as a potential cause of underestimation of sediment yield especially for soil prone to gully erosion".

3.6 Conclusion

The use of hydrological models to assess the impacts of climate may be impaired by the choice of input data. Ecosystem respond differently to the commonly use climatic variables in climate change studies. There is no agreement in the reponse of hydrological proceesses with observed and gridded data. The SWAT model is more sensitive to climate inputs than other spatial inputs. Also the sensitivity of the model to the temperature and rainfall variations due to climate change need to be established in order to make sense of results derived from climate change studies. The size of the watershed also seems to play a role in the closing of water balance within the watershed.

4. The study area

4.1 Description of the study area

The Upper Mara basin is located 250 km South West of Nairobi, covers an area of approximately 2900 km^2, and lies between 34°59'E and 35°52'E and 0°22'S and 1°13'S (Fig 4-1). The study area falls within the three counties of Nakuru, Bomet and Narok in the Rift Valley province. There are 22 divisions and 55 administrative locations in the study area (KNBS, 2009)

4.1.1 Physiography of the region

The Upper Mara River basin forms the recharge area for the Mara River basin, a key transboundary river between Kenya and Tanzania, and one of the permanent rivers feeding into Lake Victoria. The area is drained by two main rivers: the Amala and the Nyangores River which merge at the midsection to form the Mara River. The Nyangores River is the longer of the two tributaries, and has two main branches: the main Nyangores branch originates from the Mau escarpment, while the Ngetunyek branch originates from the Mau Forest. The Amala River has its origin in the Naipuipui swamp at the top of the escarpments.

According to Krhoda (2001), in this upper part of the basin the stream network are parallel pinnate, linear with numerous first order streams (106 and 102 for Nyangores and Amala respectively based on the Strahler's method of stream ordering), reflecting long parallel ridges. The streams trend in a northeast to southwest direction following the general slope of lava flow and reflect the youthful nature of the landscape. Parallel pinnate river networks are found mainly on recently formed volcanic areas. The short tributaries drain the flanks of the ridges while the long segments drain the troughs between (Krhoda, 2001).

The study area constitutes only 25% of the entire Mara River basin but is responsible for almost all the recharge and permanent flow into the 14000km^2 river basin which traverses both Kenya and Tanzania. It is delimited by the Mau escarpments to the north, and the protected reserves of the Maasai Mara/Serengeti ecosystem to the south. The area is characterized by bimodal rainfall ranging from 700 mm in the lower areas to 1800 mm in the mid and upper sections. The elevation changes from the flat tropical savanna plains at 1500 m above sea level to the high montane Mau escarpments at 3000m.

The main soil types are loams and clay loams (Andosols). The area is underlain by undifferentiated pyroclastic materials consisting mainly of poorly consolidated volcanic tuffs and volcanic ashes, which are widespread in the area and are frequently altered into clay in the upper Mau area. The most southern part of the area consists of a plain formed by deposition of subaerial volcanic ashes and later slightly dissected by shallow water courses (Mbuvi and Njeru, 1977).

According to Krhoda (2001), the local geology, topography and rainfall determine the types and distribution of soils for the Mara River Basin. The soils fit into three broad categories, namely, the mountains, plains and swamps. The mountains have rich volcanic soils suitable for intensive agricultural production including wheat, barley and zero grazing. The soils include the shallow but well-drained dark-brown volcanic soils (ando-calcaric and eutric Regosols) found on mountains and escarpments. On the hills and minor escarpments, shallow and excessively drained dark-reddish brown soils (Lithosols, mollic Andosols) are found. These soils are prone to sheet erosion and mass wasting processes and have never been cultivated before. The imperfectly drained grey-brown to dark-brown soils are found on the plateaus and high level plains. These plateaux and high plains are imperfectly drained and conducive for grass and sorghum (Krhoda (2001) The deep, dark-greyish soils (verto-eutric and Planosols) are mainly found on the Kapkimolwa plains, Shartuka and Maasai Mara National Reserve. The soils found in the study fall under "very suitable" or "suitable" classes in the suitability for agriculture nomenclature developed by Jaetzold et al. (2006).

4.1.2 Environmental status (water quality)

A number of water quality studies have been undertaken in the Mara River basin under the auspices of various actors including the Ministry of Water (Kenya), the World Wildlife Fund (WWF), the Lake Victoria Environment Management Program (LVEMP), the Nile Equatorial Lakes Subsidiary Action Plan (NELSAP) and the Global Water for Sustainability (GLOWS).

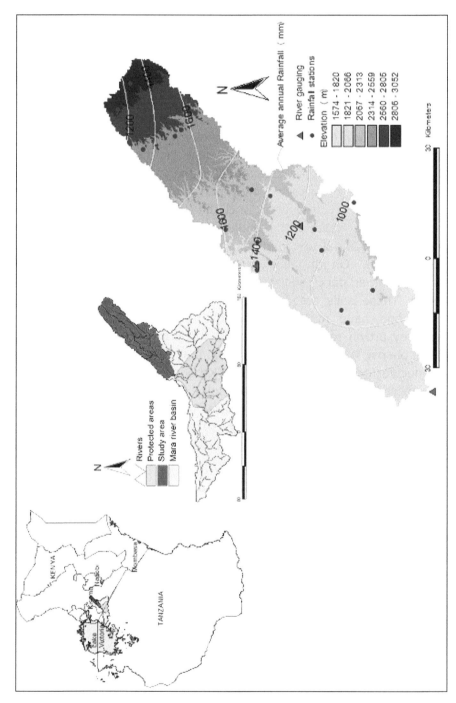

Figure 4-1: Study area as part of the larger transboundary Mara River basin.

A water quality assessment by GLOWS examined the quality of surface waters in the basin during May-2005, May-2006, and June-2007 with the goal of identifying the water quality issues and informing future monitoring and management actions (GLOWS, 2007). The assessment noted that "there is little systematic monitoring of the water quality in the basin". GLOWS also conducted three surveys of water quality at sampling stations spread across the basin from its source on the Mau Escarpment to its outlet at Lake Victoria.

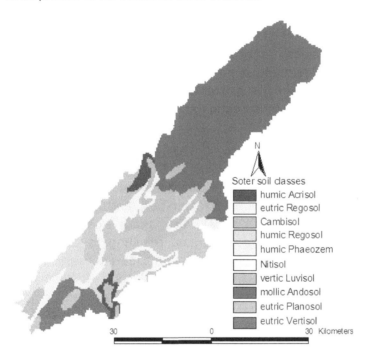

Figure 4-2: Soil map of the study area with the revised FAO 1988 naming code (adapted from Batjes and Gicheru, 2004)

The study concluded that, although no parameters were detected in excess of recognized standards, high nutrient loads and detectable amounts of mercury, pesticides, and PCBs may be impairing water quality. In recent times, courtesy of an initiative by the Lake Victoria Basin Commision (LVBC), there has been an upsurge in the number of studies aimed at understanding the spatial and temporal trends of the water quality and of the biological communities in the Mara River and its tributaries (Labatt et al., 2012; Gichana et al., 2012; Riungu et al., 2012; Mbao et al., 2012; Anyona et al., 2012 and Matano et al., 2012). Although some of these findings are preliminary and the studies are mainly focusing at the agricultural areas and are also limited in spatial coverage ($<500km^2$), it is emerging that the catchment land use has a significant influence on the water quality and on the general functioning of the river. The catchment land use has a significant influence on the functional organization of macroinvertebrate communities with shredder diversity and higher abundance in forest streams (Masese et al., 2012).

Changes in some water quality parameters between land uses, especially nitrate, chloride and sulphate, also influence macro-invertebrates, although near-stream and in-stream disturbances also play a major role (Minaya et al., 2013). Threats to the water quality in the river basin include unsewered wastewater and riparian encroachment. Stakeholders and sectors of the Mara River basin include urban settlements and villages, subsistence and large-scale agriculture, livestock, fisheries, tourism, conservation areas and biodiversity, mining and industries.

Bomet with a population of more than 80,000 inhabitants, and the other urban centres including Mulot, Olenguruone, Silibwet, and Sigor have experienced rapid population growth characterized by poor urban planning, informal settlements, limited amenities, and no wastewater treatment plants (Anyona et al., 2012). Also, there exists no formal waste handling system in the urban areas, resulting in the wanton dumping of domestic wastes along streets, in side-ditches, on river banks and into the river channel. Other potential sources of pollution include direct fetching of water in the river using donkeys and donkey drawn carts, and the direct use of river water by cattle. The use of the riparian areas for the intensive cultivation of vegetables, which requires high fertilization rates, constant irrigation and frequent chemical spraying, impacts negatively on the water quality and the ecosystem functions.

4.1.3 Demography of the region

According to the 2009 Population and Housing Census (KNBS, 2011), the population living in the three counties in which the study area lies was 3346080 persons. The study area marks the intersection of these three counties and represents only a fraction of the population. The population density is highest amongst the subsistence farming communities of the Bomet county and lowest in the herding communities of the Narok county. Bomet Township with a population of 83,729 is the largest urban centre. Other major urban centres include

Olenguruone, Mulot, Siongiroi, Chebunyo and Sigor. The area is inhabited by three main ethnic groups: the livestock herding Maasai in the lowland plains of the Narok county; the Kalenjin in both Bomet county Nakuru county, and the Kikuyu in the upstream areas in Nakuru county.

4.1.4 Socio-economics of the region

The study area falls in the Rift Valley province, the breadbasket of Kenya and home to the world reknown Maasai Mara game reserve. The Mau forest complex with the Southwest Mau forest block forming the headwaters of the basin is one of the water towers in Kenya. The main economic activities in the study area include agricultural farming, livestock herding, forest products harvesting, and tourism. Hoffmann (2007), in a study mapping water demand and use found the largest water-use factor within the MRB to be large-scale irrigation with an annual water demand of 12,323,400 m³, followed by the human population (4,820,336 m³), livestock populations (4,054,566 m³), wildlife populations (1,836,711 m³), large-scale mining (624,807 m³), and lodges and tent camps (152,634 m³).

4.1.4.1 Agriculture

There are two main types of agriculture that take place within the basin: smallholder mixed farming and large-scale commercial farming. The main cash crops grown in the area include Tea and Pyrethrum. Tea farming is carried out at large scale on tea plantations with the privately owned Kiptagich Tea factory located in the study area. Small scale tea and pyrethrum farmers with farm sizes of about two to five hectares with food crops including maize, wheat, beans and vegetables. Maize farming is spread all-over the study area, vegetable farming is restricted to the wetter upper and mid sections of the study area while wheat is predominantly grown in the drier south eastern regions of the catchment.

Due to population pressure, there is increasing encroachment of the forested areas. This has led to the excision of forest land for settlement. The agricultural expansion has also been extended to the marginal areas outside the Agro-Ecological zoning (AEZ). According to George and Petri, (2006), AEZ provides a standardized framework for characterizing 3 major resources relevant to agricultural production, namely the prevailing climate (temperature, water and solar radiation), soil (texture, drainage, depth and stoniness) and terrain conditions (slope, aspect, configuration, micro-relief). The methodology identifies the suitability of specified land uses under assumed levels of inputs and management conditions. A comparison of a suitability map developed from the AEZ maps for Kenya (Jaetzold et al., 2003; Jaetzold et al., 2006) with the FAO landcover map of Kenya (Fig. 4-3) shows the cultivation of crops in areas previously defined as unsuitable for their cultivation.

4.1.4.2 Forest

The four forest blocks making up the Mau forest complex from the west are; the Western Mau, the Southwest Mau, the East Mau and the Maasai Mau which are significant due to their role as water catchment. Southwest Mau forest, being at 3000m above mean sea level, constitutes the uppermost part of the Mara River Basin. The Southwest Mau Forest cover helps infiltration and percolation of rainwater into the ground, and is therefore the source of Mara's main tributaries Nyangores and Amala (Krhoda, 2001). According to Kinyanjui, the afromontane mixed forest the varies in species composition and experiences varying levels of human degradation.

The forests on the windward side of Lake Victoria are mainly moist mixed forests dominated by *Tabernamontana – Allophylus – Drypetes* forest formations while the Eastern Mau is dominated by a dry upland conifer forest dominated by *Juniperus procera* (Hochst.ex Endl.) and *Podocarpus latifolius* (Thunb. Mirb). Primary colonizers like *Neoboutonia macrocalyx* (Pax), *Macaranga kilimandischarica* (Pax) and *Dombeya torrida* (J.F. Gmel.) are characteristic for degraded forests (Kinyanjui, 2010). The forest provides honey, fuelwood, medicinal herbs, forest employment and forest farming. The greatest threat of the forest is both legal and illegal excisions and over-exploitation. Illegal logging in the forest, charcoal burning and uncontrolled livestock herding in the forest pose an immediate threat to the forest ecosystem.

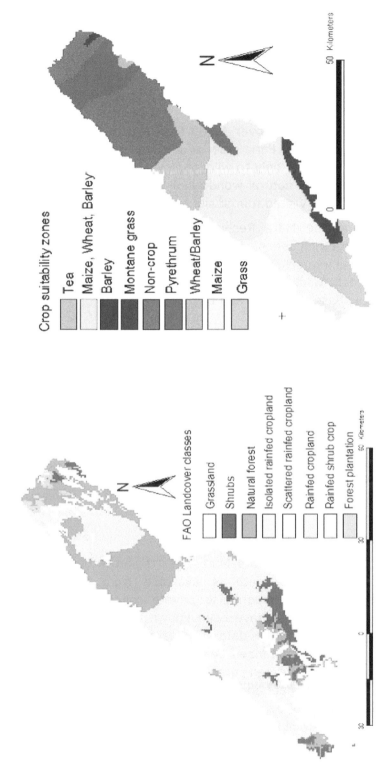

Crop suitability zones

Tea
Maize, Wheat, Barley
Barley
Montane grass
Non-crop
Pyrethrum
Wheat/Barley
Maize
Grass

N

50 Kilometers

FAO Landcover classes

Grassland
Shrubs
Natural forest
Isolated rainfed cropland
Scattered rainfed cropland
Rainfed cropland
Rainfed shrub crop
Forest plantation

N

Figure 4-3: Landcover classifications in the study area (Di Gregorio and Jansen, 2000) and suitability zonations for different crops (Jaetzold et aL, 2006).

4.1.4.3 Tourism

The Mara River, transboundary between Kenya and Tanzania, plays a major role in the tourism industry by being the only water source for the Maasai Mara Game Reserve (Kenya) and the upper parts of the Serengeti National Park (Tanzania) during the dry season. The Maasai Mara National Reserve covering over 1,500 km^2 and an extension of Tanzania's Serengeti National Park reserve are gazetted consrvation areas, with land use restricted only to wildlife viewing tourism. In 2006, it was voted as one of the 7 natural wonders of the world and is one of the most famous and most visited tourist sites in Kenya.

The land sorrounding the Mara Reserve is managed by groups of pastoralists pooled together under umbrella organisations known as co-operative group ranching schemes. These areas contain year-round communities of resident wildlife, but migratory wildlife also spill out onto them during the dry season (Seernel and Lambin, 2001).

4. 2 Data used for the research

Different data sets, accessed from various sources (local, regional and global) have been used through this study. The data used for the research included historical observation data for climatic and streamflow variables, data collected through field surveys, and satellite remote sensing data. In order to make the data useable in the various applications in the study, several interpretations, processing and manipulation procedures were performed using simple and/or complex algorithms. Both open source and proprietary softwares and tools were deployed. A brief description of the open/freeware tools used is given in Annex 1.

4.2.1 Hydrometeorologic data

Generally hydro-climatic data for the Mara river region is both scarce and scanty. Officially rainfall data is available from the Kenya Meteorological department, although also private concerns (including hotels, lodges, and tea estates) have been known to collect and keep especially rainfall and temperature data. The data scarcity problem in the study area has been characterized by Gann (2006): "data are non-existence, non accessible, limited accessible, with gaps, not very accurate, and in different formats and no standards". The problem is not limited to the Mara basin only, but is widespread across the Lake Victoria region and the African continent as a whole (Ndomba et al.,

2008; Kingston and Taylor, 2010). To improve on the data quality, and to overcome some of the challenges in the wider Lake Victoria region, various researchers have used different manipulations techniques. Mati et al., (2008) e.g. used the inverse distance weighting (IDW) technique for rainfall and the long term daily mean for the flow. Ndomba et al. (2008) also used the long term daily mean for the flow. Others, including Kingston and Taylor (2010) and Mango et al. (2011), have resorted to using remotely sensed satellite climatic data. The replacement of locally observed data with global databases has some shortcomings. Koutsouris et al. (2010) in a study of a small catchment near Lake Victoria found clear disparity between discharge observations at the considered regional and local scales, leading to different hydrological trend assessments based on the data from the different scales.

4.2.1.1 Rainfall data

Three air masses influence the rainfall regime of the Mara River Basin. The apparent movement of the Inter-Tropical Convergence Zone (ITCZ) determines the seasons (Krhoda, 2001). The catchment is dominated by dry northeasterly winds from the Sahara Desert from November through March causing little rainfall. Short rains are experienced from November to December. The Southeast Trade winds from the Indian Ocean influence the rainfall pattern of the region between March and June, weakening considerably between June and October. Compared to August and September, the less dry months are January and February. The southwest trade winds, or sometimes known as the Congo air mass, bring rain from the west in July with storms and hailstorms (Krhoda, 2001). Three climatic zones, namely semi arid, sub humid and humid are defined. Southwards the humid zone two rainfall peaks occur between March and May and another peak between July and September. The sub-humid zone also has two rainfall peaks, March-May and July-September. In the semi arid zone, the rainfall has a monomodal pattern with a dry season in the months of June to November. These rainfall patterns are demonstrated in Fig. 4-4 for the Baraget, Koiwa, and Kabason stations representing respectively, the humid, sub-humid and semi-arid regions.

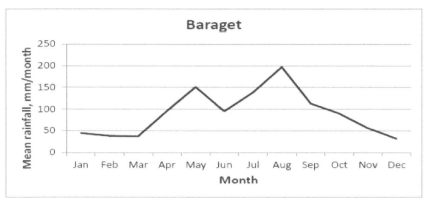

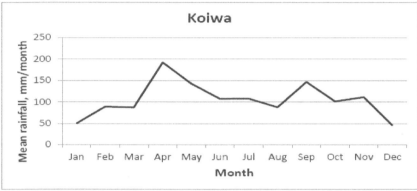

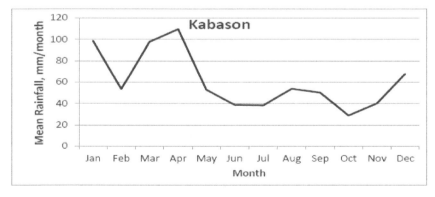

Figure 4-4:Typical annual rainfall patterns for humid (Baraget), sub-humid (Koiwa) and semi-arid (Kabason) climatic zones in the region

Comparison of the rainfall volumes for the upstream and downstream parts of the basins has shown that the mountainous and high elevation areas of the upstream watersheds receive more rainfall than the lower portion of the basin, with inter-stations variability (Melesse, 2008). There are eighteen (18) rain gauging stations within the study area (Fig.

4-5). The length of observations and the percentage of data availability over that period (% filled) for the rainfall gaging stations is given in table 4.1. None of these is a World Meteorological Organization (WMO) climatic station, although three WMO stations are available within a radius of 100 km from the basin. Rainfall, temperature, wind speed, relative humidity and solar radiation were available from two additional synoptic stations situated in the vicinity of the study area, and managed by the Kenya Meteorology department. Global climatic databases including the University of East Anglia, the CRU2.0 database (New et al. 2002), and the FAO AQUASTAT database (FAO, 2005) have also been used for evapotranspiration and other meteorological data.

Tabel 4-1. Rainfall gauging stations within the Upper Mara River basin

ID	Name	Lat.	Long.	Elevation	Start	End	% filled
9035079	Tenwek Hospital	-0.750	35.333	2012	1960	2002	91
9035085	Olenguruone DO	-0.583	35.683	2743	1960	2004	80
9035126*	PBK Bomet	-0.783	35.333	1981			
9035227	Bomet DC	-0.783	35.333	1951	1960	1992	94
9035228	Kiptunga forest	-0.450	35.800	1829	1961	2009	94
9035241	Baraget forest	-0.417	35.733	2865	1961	1999	93
9035265	Bomet water	-0.783	35.350	1920	1967	2009	88
9035324	Keringet forest	-0.483	35.633	2560			
9035284	Mulot police	-0.933	35.433		1973	1997	96
9035302	Nyangores forest	-0.700	35.433	2219	1980	2009	96
9035339*	Kiptagich farm	-0.567	35.667	2341			
9135001	Narok	-1.13	35.83	1890	1960	2010	99
9135008	Kabason AGC	-1.000	35.233	1646	1960	1986	92
9135010	Aitong	-1.183	35.250	1829	1960	1992	54
9135019	Lemek maasai	-1.100	35.38	1829	1966	1993	71
9035260	Koiwa estate	-0.817	35.350	1916			
9035312*	Merigi chiefs centre	-0.783	35.400	2134			
9135027*	Emarti Health centre	-1.017	35.200	1768			
9035334*	Sogoo health centre	-0.833	35.600	2134			
*	Stations either under private ownership or newly established						

4.2.1.2 Streamflow data

River flow data is collected and managed by the Water Resources Management Authority (WRMA), and its parent Ministry of Water and Irrigation. There are three established river gauging stations (RGS) in the upper Mara basin. 1LA03, situated on the Nyangores tributary, 1LB02, on the Amala, and RGS 1LA04 situated downstream of the confluence between the Amala and the Nyangores (Fig 4-5). 1LB02 is located along a bridge abutment at Kapkimolwa, hence the cross-sectional area is stable. 1LA03 is located at Bomet, on a section that is prone to bank erosion and sedimentation. The original 1LA4 manual staff gauge was located on a hippotamus prone area making maintenance impossible. The 1LA04 has since 2010 been replaced with an automatic river gauging station (Khisa, 2012). Because of the source the Nyangores and Amala rivers is in the forest region, they experience low evaporation loss and high rainfall contributing to the normal discharges of the Mara River. The details of the flow gauging stations are provided in Table 4-2, while the flow characteristics for the period 1970-1977 is given by the single mass curves (Fig 4-6)

Tabel 4-2. River gauging stations on the entire Mara River

ID	Code	River	Location	Lat.	Long.	Start	End	% filled
	1LA01	Nyangores	Nyangores	-0.072	35.358	No records available for these river gauging stations		
	1LA02	Nyangores	Keringet	-0.265	35.688			
	1LB01	Amala	Awaja	-0.280	35.416			
107052	1LB02	Amala	Kapkimolwa Bridge	-0.897	35.438	1955	1995	
107032	1LA03	Nyangores	Bomet bridge	-0.786	35..255	1963	1992	87
107062	1LB04	Mara	Lalgorian Narok Rd	-1.267	35.017	1970	1992	
	1LA05	Mara	Serena					
107072		Mara	Mara Mines	-1.65	34.564	1970	1977	76
107081		Mara	Mara Ferry	-1.524	33.975	1970	1978	-
109012	In TZ	Mara	Mara mines	-1.769	33.683	1970	1992	-

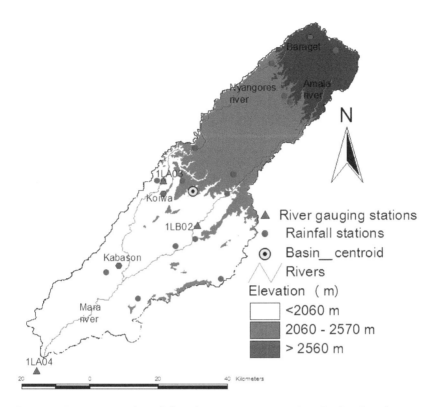

Figure 4-5: Location of rainfall and river gauging stations in the river basin

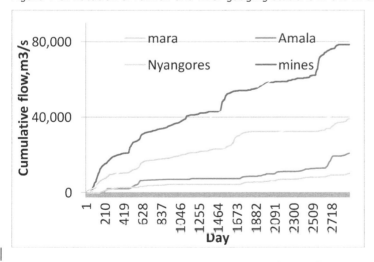

Figure 4-6. Single mass curves for discharge at different flow gauging stations along the Mara River.

4.2.2 Global databases

4.2.2.1 The digital elevation model (DEM)

Burrough, and McDonnell, 1998 defined digital elevation model (DEM) as "any digital representation of the continuous variation of relief over space". It is a regular two dimensional array of heights sampled above some datum that describes a surface, and contains elevation information with the addition of some explicit coding of the surface characteristics such as breaks in slope or drainage divides (Wood, 1996).

The two sets of DEM databases with a 80% world coverage are commonly used: the Advanced Spaceborne Thermal Emission and Reflection Radiometer (ASTER) DEM and the Shuttle radar topography mission (SRTM) (Gesch, 2012). AsterDEM has a finer resolution of 15-30m while the SRTM has a 30-90m resolution. Comparative studies conducted by Nikolakopoulos et al. (2006) showed a strong correlation between the two datasets. While comparing the effect of DEM resolution on hydrological modeling, Tulu (2005) found no significant difference in the monthly runoff when the AsterDEM is replaced with the SRTM DEM.

The 3 arc sec horizontal resolution SRTM DEM (Farr and kobrick, 2000) was used in this study. The key characteristics of the SRTM DEM include: 16-bit signed integer data in a simple binary raster, data provided with no embedded header or trailer and stored in row major order, elevations referenced to the WGS84/EGM96 geoid and range from -32767 to 32767 meters, Motorola "big-endian" byte order standard with the most significant byte first used. The SRTM3 DEM's have a resolution of 3 arc second (90m) at the equator, and are provided in mosaiced 5 deg x 5 degree tiles for ease of download and use. Digital elevation models (DEM) tiles for the Mara river basin (S01E034, S02E034, S01E035, and S02E035) were downloaded from the USGS server http://dds.cr.usgs.gov/srtm/version2_1/SRTM3/Africa/.

The SRTM files are accessed in the Geoid Height File (hgt) format, and were converted to Geographic Tagged Image File Format (Geotiff) using the 3DEM software (Horne, 2009). Swapping of the bytes from the most significant byte first ("big endian") to the least significant byte first ("little endian") was performed using the VT builder software (Discoe, 2009). The DEM was also visualized to ascertain that data properties that might otherwise render the data unsuitable for the planned application are fulfilled including the effect of swapping byte order.

4.2.2.2 Soil data

The KenSOTER database (Batjes and Gicheru, 2004) was used for the soil information. According to Batjes and Gicheru, (2004), the land surface of the Republic of Kenya excluding lakes and towns has been characterized using 397 unique SOTER units corresponding with 623 soil components. The major soils have been described using 495 profiles, which include 178 synthetic profiles, selected by national soil experts as being representative for these units. The associated soil analytical data have been derived from soil survey reports and expert knowledge. Gaps in the measured soil profile data have been filled using a step-wise procedure which includes three main stages: (1) collate additional measured soil data where available; (2) fill gaps using expert knowledge and common sense;and (3) fill the remaining gaps using a scheme of taxotransfer rules.

Twenty one (21) soil classes were identified to belong to the study area and reclassified to the revised FAO (1988) and the Kenya soil survey (KSS) legends. The KSS legend was adopted for SWAT because unlike the FAO legend, there were no overlaps in the soil names. The proportions of the soil type in the study area and the corresponding FAO and KSS labels are given in Annex 2.

4.2.2.3 Landsat images

Landsat images for the area of interest (0°-1°S, 35°-37°E) were obtained from Global Land Cover Facility (http://glcfapp.umiacs.umd.edu:8080 /esdi/index.jsp) and USGS Global Visualization (http://glovis.usgs.gov/). For imagery to be used in the land use change analysis, in order to allow for consistency in the classification process, images of the same season of the year, also referred to as "anniversary dates" (late January- early February) were used (Table 4-1). In addition, Landsat images of February and October 2006 were also used for additional information on the development of a landcover map. The anniversary datasets served as the primary Landsat dataset while the additional images were used as supplimentary data.

Table 4-1: Landsat imagery used for land use mapping and landcover change analysis.

Year	Landsat Instrument	Image tile	Acquisition date
1976	Multispectral Scanner System (MSS)	181/60	12/02/1976
		181/61	12/02/1976
		182/60	13/02/1976
		182/61	13/02/1976
1986	Thematic Mapper (TM)	169/60	28/01/1986
	Thematic Mapper (TM)	169/61	28/01/1986
1995	Thematic Mapper (TM)	169/60	06/02/1995
	Thematic Mapper (TM)	169/61	06/02/1995
2006	Enhanced Thematic Mapper Plus (ETM+)	169/60	27/01/2006
		169/61	27/01/2006

4.2.2.4 Normalized Difference Vegetation Index (NDVI) images

The NDVI, first proposed by Rouse et al. (1973), is the most commonly used and readily available vegetation index with long term archives. NDVI measures the amount of energy absorbed by leaf pigments such as chlorophyll, and is closely correlated with the fraction of photosynthetically active radiation (fPAR) absorbed by plant canopies and therefore leaf area, leaf biomass, and potential photosynthesis.

$$NDVI = \frac{NIR-IR}{NIR+IR} \text{---4-1}$$

Where: NIR and IR are spectral reflectance measurements acquired in the near-infrared and visible (red) regions

Different preprocessed NDVI datasets having different spatial and temporal resolutions, and with different temporal coverage are available from different sensors. The main characteristics of commonly used, freely available NDVI datasets are given in Table 4-2.

Table 4-2: The different NDVI datasets (Pettorelli et al 2005).

Dataset	Satelite	Instrument	Temporal span	Temp resol.	Missing data	Range
PAL	NOAA	AVHRR	July 1981-Sep2001	1 day, 10-day	sep 1994-jan 1995	8km
GVI	NOAA	AVHRR	may 1982-present	weeklymonthly	sep 1994-jan 1995	16km
GIMMS	NOAA	AVHRR	July 1981-presdt	bi-montly	None	8 km
MOD1	TERRA	MODIS	Feb 2000-present	16 days	None	250-1km
	TM/ETM	Landsat	1984-present	16 days	None	<10-30m
	VGT	SPOT	April 1988-present	10 day	None	1km

34

The Satellite Pour l'Observation de la Terre-VEGETATION (SPOT_VEGETATION, abbreviated as SPOT_VGT) database was used in this study. According to Toukiloglou (2007) in a study to compare between the Moderate Resolution Imaging Spectroradiometer (MODIS1), Advanced Very High Resolution Radiometer (AVHRR) and VEGETATION datasets, the VEGETATION datasets have produced significantly more accurate land cover maps in most of the cases (five out of eight). They attributed the better performance of the VEGETATION datasets over MODIS1 to the additional spectral bands they contained. The datasets of VEGETATION had a lower spatial and radiometric resolution than MODIS1 as well as wider spectral bands. The need to link recent NDVI archives (MODIS, SPOT) to the longer series AVHRR in order to have longer and more reliable time series have been made difficult by the coarse resolution (8 km) AVHRR data that has been shown to miss broad scale changes in vegetation cover (Yin et al., 2012).

The NDVI data from SPOT-VGT ("VGT-S10" product) were downloaded from VITO (Flemish Institute for Technological Research; VITO 2006). VGT-S10 ten day maximum value composite (MVC) synthesis data are a series of data segments that have been acquired in a ten days period. To ensure good quality of the MVC data pixel by pixel comparisons are performed on the segments of the period of interest. MVC syntheses having a spatial resolution of 1km*1km for 13 yrs (04/1998- 05/2011) were accessed. MVC provides a method for improving the accuracy of green vegetation monitoring with NDVI. By combining daily reflectance images over a specific time period into a single NDVI image, MVC minimizes cloud effects, reduces view angle effects and mitigates atmospheric water-vapor and aerosol contamination, thus separating the green vegetation land cover from various other components.

The intensity of a pixel is digitised and recorded as a digital number. Due to the finite storage capacity, a digital number is stored with a finite number of bits (binary digits). The number of bits determines the radiometric resolution of the image. (i.e 8-bit digital number ranges from 0 to 255, while an 11-bit digital number ranges from 0 to 2047).

The digital number (DN) is calculate from the real NDVI using the expression:

$$\text{Real NDVI} = a * DN + b \text{ ---}4.2$$

Where: a and b are coefficients;

4.2.2.5 Leaf Area Index (LAI) images

The leaf area index (LAI) is the one sided area of green elements (defined by a chlorophyll content higher than 15µg.cm-²) per unit leaf horizontal soil, and is strongly non linearly related to reflectance. The LAI represents the quantity of foliage in the pixel area. Basically, LAI=0 corresponds to bare soil; LAI=5 or 6 characterizes a dense canopy (Baret et al. 2004). It can be related to processes such as photosynthesis, evaporation and transpiration, rainfall interception and carbon flux. Long-term monitoring of LAI can provide an understanding of dynamic changes in productivity and climate impacts on forest ecosystems (Zheng and Moskal 2009). Further, since LAI remains consistent while the spatial resolution changes, estimating LAI from remote sensing allows for a meaningful biophysical parameter, and a convenient and ecologically-relevant variable for multi-scale multi-temporal research that ranges from leaf, to landscape, to regional scales.

According to Zheng and Moskal (2009), there are two types of methods for the estimation of LAI: either employing the "direct" measures involving destructive sampling, litter fall collection, or point quadrat sampling "indirect" methods involving optical instruments and radiative transfer models. The dynamic, rapid and large spatial coverage advantages of remote sensing techniques, which overcome the labor-intensive and time-consuming defect of direct ground-based field measurements, allow remotely sensed imagery to successfully estimate biophysical and structural information of forest ecosystems. Ground-based measurements have no standards as several methods, like harvesting methods, hemispherical photography or light transmission through canopies, can be used. LAI maps generated at various spatial resolutions from a daily to monthly period over the globe using optical space borne sensors can therefore be used to complement ground LAI. Freely available LAI products have been developed including: Advanced Very High Resolution Radiometer (NOAA-AVHRR), Moderate Resolution Imaging Spectroradiometer (MODIS), Spinning Enhanced Visible and Infrared Imager (SEVIRI) and Satellite Pour l'Observation de la Terre-VEGETATION (SPOT-VGT) data.

The effective LAI derived from the SPOT-VGT Carbon Cycle and Change in Land Observational Products from an Ensemble of Satellites (CYCLOPES) processing chains at a 10-day temporal frequency supplied by the VGT4Africa project (Baret et al., 2006) was used in this study.

The CYCLOPES LAI time series used in this study is based on daily observations of the SPOT-VGT sensor (Gessner et al. 2013). Preprocessing of VGT data includes radiometric calibration, cloud screening, atmospheric correction and BRDF normalization. CYCLOPES LAI data is available at a spatial resolution of 1 km. According to Gessner et al., 2013, the SPOT-VGT was chosen over the MODIS dataset since MOD15A2 shows deficits with regards to the smoothness of temporal profiles in the more humid zones, and that the SPOT-VGT has superior ability to reproduce vegetation phenology. Further, frequent residual cloud contamination in the study area seems to be a particular problem of the MODIS dataset.

4.3 Conclusion

Although rainfall data is the only recorded hydrometeorological parameter within the station, there is data available from other sources within and outside the Lake region. Global databases have been accessed from and courtesy of various sources. All the global databases used were freely available for download either directly or through a registration   process.

5. Remote sensing data analysis

5.1 Introduction

Campell, 2002 defined remote sensing as "the practice of deriving information about the earth's land and water surfaces using images acquired from an overhead perspective, using electromagnetic radiation in one or more regions of the electromagnetic spectrum, reflected or emitted from the earth's surface". Remote sensors can be deployed on satellites, airplanes, balloons, or remote-controlled vehicles. Satellite images offer unique advantages over other sources of spatial-temporal information. These include: large area coverage, cost effectiveness, time efficient, multi-temporal, multi-sensor multi-spectral and overcomes inaccessibility and allows faster extraction of GIS-ready data.

Remote sensing offers unique advantages to other sources of data. These include; provision of map-like representation of the Earth's surface that is spatially continuous and highly consistent, repetitive data acquisition, availability at different ranges of spatial and temporal scales, digital formats (Lu et al., 2001; Foody, 2002). The main shortcomings of satellite images include the need for both ground-truthing and for an expert system to extract data. In order to make the output of this study applicable to a wide audience, and encourage reproducibility of results, freely available and easily downloadable satellite images of varying resolutions were used to achieve the set goals.

There are numerous classification algorithms and approaches which may be broadly categorized as manual, automated and hybrid. Manual classification relies on the interpreter to employ visual cues such as tone, texture, shape, pattern and relationship to other objects to identify the different land cover classes and is effective when the analyst is familiar with the area being classified (Horning et al., 2010). The manual classification is limited to only colour images, comprising red, green and blue bands, and does not take the advantages presented by multi-spectral images into account. It is tedious, time consuming and subjective.

Automated classification uses an algorithm to generate and apply specific rules to assign pixels to one class or another. Automated algorithms incorporate hyperspectral as well as hypertemporal layers of satellite data, along with assorted ancillary data layers. Pixel procedures analyze the spectral properties of every pixel within the area of interest, without taking into account the spatial or contextual information related to the pixel of interest. Pixel based methods rely either on supervised classification, on unsupervised classification or on a combination of both. In supervised classification, sample pixels (known as signatures or training sites) are identified and used as representative examples for a particular land cover category and the sample pixels are then used to train the algorithm to locate and classify similar pixels in the image.

In the unsupervised classification, the algorithm groups pixels together into unlabeled clusters and then have the analyst label the clusters with the appropriate land cover category. The difference between the different types of supervised classification algorithms is how they determine statistical similarity between pixels (Horning et al., 2010). Common supervised statistical classification algorithms include; minimum distance, mahalanobis distance, maximum likelihood and parallelepiped methods (Jensen, 2005). Other algorithms include artificial neural networks (ANN) which mimic the human learning process or the binary decision trees that use a set of binary rules.

The biggest setback with per pixel classification is the salt and pepper effect (Jensen 2005). The salt and pepper noise (SPN) contains random occurrences of both black and white intensity values, and is often caused by the threshold of noise in an image (Al-amri, 2010), where many single pixels of a particular class exist that are interspersed with contiguous areas of other classes (Knight and Lunetta, 2003). To address this challenge, object-based sub-pixel classification procedures have been developed. These packages analyze both the spectral and spatial/contextual properties of pixels and use a segmentation process and iterative learning algorithm to achieve a semi-automatic classification procedure that promises to be more accurate than traditional pixel-based methods (Riggan and Weih , 2009). Object orientation or segmentation involves the comparison of a pixel's value with values of the neighboring pixels. If neighboring pixels are similar, they are added to the contiguous group and if they are not, then another segment is started.

For this study, automated statistical classification methods were used on low resolution (1km to 10 km SPOT data) to medium resolution (10-100 m Landsat images) data. All the satellite images used in this study were freely available for download by the public courtesy of the custodians from their respective repositories. Landsat data included multispectral images from the multispectral scanner (MSS), Thematic Mapper (TM) and Enhanced Thematic Mapper Plus (ETM+) instruments. They were accessed from the Glovis webpage (http://glovis.usgs.gov/). The SPOT_VEGETATION (SPOT-VGT) sensor onboard the SPOT 4/5 instrument was used for multitemporal data on Normalized Difference Vegetation Index (NDVI) and Leaf area index (LAI) data, and was accessed from the Flemish Institute for Technological Research (VITO)(http://free.vgt.vito.be). Both multi-temporal and multispectral satellite images were used to develop classified land use/land cover maps, while only multi-spectral images were used to analyse historical long-term land cover change dynamics.

5.2 The development of classified landcover maps based on Landsat images

To adequately cover the entire study area, four scenes for the multispectral scanner (MSS), and two for the TM and ETM+ Landsat sensors were used. Table 4-1 summarizes the characteristics of the Landsat images used in this study. The images were pre-processed, processed, and post-processed to obtain the classified map. Pre-processesing included geo-referencing of images, layerstacking of bands, mosaicking of tiles, and subsetting of the region of interest. Processing steps included both unsupervised and supervised classification. Post classification processes included accuracy assessment and majority filtering.

The proprietary ERDAS IMAGINE 9.2 (Leica geosystems) image processing software was used in the classification process. The Anderson system (1976) and the FAO/UNEP Land Cover Classification System (LCCS) FAO, (2005) were adopted for the classification schemes. The purpose of such schemes is to provide a framework for organizing and categorizing the information that can be extracted from the data, since the proper classification scheme includes classes that are both important to the study and discernible from the data on hand.

5.2.1 Preprocessing

In order to analyse remotely sensed images, the different images representing different bands must be stacked. This allows for different combinations of Red Green Blue (RGB) to be shown in the view. Landsat images from MSS are in 6 bands, while those from TM and ETM+ are in 8 bands. Layerstacking was performed to combine all the image bands minus the thermal bands. The study area spanned several image files.

Image mosaicking which involved the combination of the two TM/ETM+ and four MSS images was performed to create one large file.

An Arcview shapefile for the watershed, geo-referenced to the same coordinate system as the mosaiced image, was used to get a subset of the images for the catchment. Sub-setting not only eliminates the extraneous data in the file, but it speeds up processing due to the smaller amount of data to process, which is important when dealing with multiband data ERDAS, (2005). The extracted subset (Fig. 5-1) was used in the classification procedures of the images. Unsupervised classification followed by supervised classification was used. According to ERDAS (2005), combining supervised and unsupervised classification may yield optimum results, especially with large data sets.

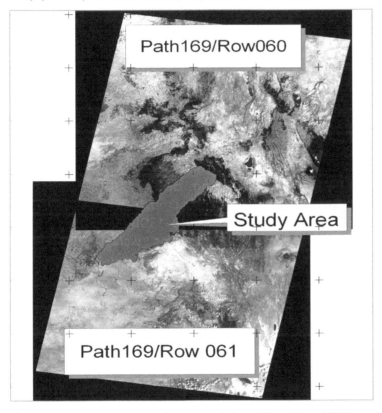

Figure 5-1. Coverage range of mosaiced tiles of the TM and ETM+ Landsat sensor

5.2.2 Unsupervised classification

The Iterative Self-Organizing Data Analysis Technique (ISODATA) clustering method was used for unsupervised classification (Tou and Gonzalez, 1974). ISODATA uses spectral distance, but iteratively classifies the pixels, redefines the criteria for each class, and classifies again, so that the spectral distance patterns in the data gradually emerge. It uses minimum spectral distance to assign a cluster for each candidate pixel. The process begins with a specified number of arbitrary cluster means or the means of existing signatures, and then it processes repetitively, so that those means shift to the means of the clusters in the data (ERDAS, 2005).

To perform ISODATA clustering, the following parameters were specified: the maximum number of clusters to be considered was set at 20; a 95% convergence threshold, and the maximum number of iterations to be performed was set to 100. Due to the coarseness of the images (30m x 30m resolution) and the limitation of extensive field data, a generalised hierarchical classification scheme system was used in this study. The terminology used for the classes was consistent with the one used in the FAO/UNEP international Land Cover Classification System (LCCS) standard (FAO, 2005). The unsupervised classes were further recoded to six (6) clusters including: closed to open trees, trees or shrubs, closed to open shrubs, herbaceous crop, grassland and mixed vegetation cover. During the recoding of the ISODATA clusters, new signatures were generated. These signatures were used in the supervised classification.

5.2.3 Supervised classification

Remote sensing data acquired in the dry season has a less spectral response from the vegetation in the agricultural area as most of the area is bare. This situation makes it easy to separate natural vegetation from managed cultivated area by use of the dry period Landsat image of Jan/Feb period. Due to the difficulty in differentiating cropped area and areas under grass (which also show poor reflectance in the dry season), two more Landsat images for June and October were used to generate additional signatures. June represents the peak growing season for all vegetation in the study area, whereas in October, grass with a shorter growing period will have dried out (green off) leaving the crops still growing (green on).

The signatures from the six classes developed during the unsupervised classification (§5.2.2) were used with the Minimum-Distance to the Mean classifier method to perform the initial supervised classification. The second stage of the supervised classification was conducted using the parallelepiped classifier method with the aid of training sites acquired on known land cover types. Training sites used were those of known land cover types like tea plantations, irrigated agricultural fields and forest. The known spatial information of the selected sites was used to generate non-parametric signatures. Parametric training sites were also generated by use of spectral reference points at specific geographical location. Ground truth for these training sites was acquired through field transect surveys, analysis of aerial photography, secondary maps and local expert knowledge.

Level-1 land cover maps, as recommended for Landsat images in the Anderson Classification system (Anderson, 1976), featuring four Land Cover classes were developed. The Class definitions used included: forest (closed to open trees), shrubland (trees or shrubland, closed to open shrubland), cropland (herbaceous crop) and grassland (savanna grassland and scattered grass/bareland areas). As is often characteristic of pixel-by-pixel classifiers, the map suffered from the "salt and pepper effect" (Lillesand and Kiefer, 1987). According to Blaschke, 2000, in the per-pixel characterization of land cover a substantial proportion of the signal apparently coming from the land area represented by a pixel comes from the surrounding pixels. This is the consequence of many factors, including the optics of the instrument, the detector and the electronics, as well as atmospheric effects.

5.2.4 Post classification

To reduce the "salt and pepper effect" and minimize registration problems, a majority filter with a 3 x 3 pixel square was used. A single pixel is a poor sample unit since it is an arbitrary delineation of the land cover and may have little relation to the actual land cover delineation. Further, it is nearly impossible to align one pixel in an image to the exact same area in the reference data. In many cases involving single pixel accuracy assessment, the positional accuracy of the data dictates a very low thematic accuracy (Congalton, 2005). Figure 5-2 presents the thematic land cover maps obtained from the classification process.

5.3 The development of classified landuse map based on SPOT-VGT NDVI

A total of 474 S10 NDVI images were extracted with a polygon within the bounds 0° to 2° south, and 34° to 36° east from the NDVI_ AFRICA region, using the VGTExtract tool provided by VITO. The extracted images were layered together to make 4 single images, each covering 3 years (thus 04/1998-03/2001, 04/2001-03/2004, 04/2004-03/2007, 04/2007-03/2010). The subsets covering only the study area were derived from the single image with 108 decades (i.e. a total 1095 days). The layer-stacking and sub-setting was performed using the ERDAS Imagine 9.2 image processing software. The series between 04/2004 and 03/2007 was used for the classification, this was to enable the use of the Landsat images of 2006 which had the lowest cloud cover noise interference. The resulting Landsat derived map served as a baseline map for the NDVI mapping to provide a mask for the naturally vegetated areas.

The ISODATA technique was used to perform unsupervised classification of the NDVI data. The algorithm splits and merges clusters, in which cluster centers are randomly placed and pixels are assigned based on the shortest distance to the center method. The program runs the algorithm through many iterations until the user defined convergence threshold or the specified number of iteration runs is reached. The optimum number of clusters was determined by examining the separability of the clusters. Choosing between 10 and 50 clusters, the ISODATA algorithm was ran repeatedly for each cluster number for 100 iterations and divergent convergence threshold of 1.0. The best number of classes to be chosen was based on the maximum values of the average and minimum divergence statistics with a clear and distinct peak in the separability values of divergence. A total of 30 clusters were chosen for this study. The classes were further refined by the use of the transformed divergence measure.

In the transformed divergence separability the range of divergence varies between 0 and 2000, with 0 indicating complete overlap between the signatures of two classes, and 2000 indicates a complete separation between the two classes.The larger the separability values are, the more likely the final classification results will be good. Swain and Davis (1978) defined values greater than 1900 as separable, 1700-1900 as weak separable and <1700 as inseparable.

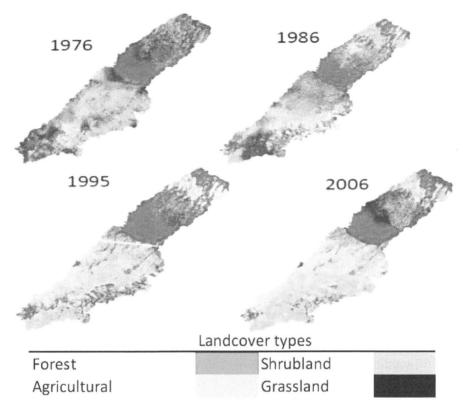

Figure 5-2: The land cover maps based on Landsat images (MSS for 1976, TM for 1986 & 1995 and ETM+ for 2006)

Using this rule, the classes were merged from the initial 30 classes to 23 classes. Once the final classes where identified, the time series of the NDVI data were extracted. The 108 decades represent a continuous stream of phenological information spanning over 3 years. Changes in NDVI time-series indicate changes in vegetation conditions proportional to the absorption of photosynthetically active radiation (PAR). However, there are nearly always disturbances in these time series, caused by cloud contamination, atmospheric variability, and bi-directional effects. These disturbances greatly affect the monitoring of land cover and terrestrial ecosystems and show up as undesirable noise (Chen et al., 2004). The most widely used techniques developed to eliminate noise in NDVI time series caused by clouds, ozone, dust, as well as off-nadir viewing and low sun zenith angles include: Best Index Slope Extraction

(BISE), Fourier based filtering, Savitzky-Golay, asymmetric Gaussian, and logistic function filters (Yin et al., 2012).

The NDVI curves were smoothened with the fitting algorithms in the TIMESAT tool (Eklundh and Jönsson, 2010), a software package for analyzing time-series of satellite sensor data. Output from the TIMESAT program is a set of files containing seasonality parameters; beginning of season, end of season, amplitude, integrated values, derivatives, etc., as well as fitted function files containing smooth renditions of the original series (Eklundh and Jönsson, 2010). The Savitzky–Golay filter (Savitzky and Golay, 1964) was used for single season profiles, while the Gaussian or logistic filters were used for the multi-season profiles. Seasonality parameters, including the beginning of the season, the end of the season, the length of the growing period and the amplitude were also extracted.

Three categories of land cover classes were identified, namely those with distinct flat/plateau profiles, those with clear sinusoidal phenological profiles and those with mixed patterns. Using the land cover map obtained from the Landsat imagery (§5.2), the areas under perennial land cover were masked out. The phenological profiles in the masked area (Fig. 5-3a) have plateau shaped vegetation cycles and NDVI values that remain relatively high (~200) over extended periods of time. These represent areas with a more permanent cover of natural vegetation like forests, tree plantations and shrub crops. The remaining two categories are affected by climatic conditions, they exhibit a rainfall mediated NDVI cycle with a Gaussian distribution shape. Degraded areas, represented by herbaceous vegetation like annually cropped areas and grass cover areas, exhibit a distinct seasonal response driven by rainfall incidences and reduced NDVI values, especially during the established dry period (December-February) (Fig. 5-3b). The rest of the areas have no proper rainfall seasonal pattern but exhibit clear minima and maxima that allow separation of phenometric parameters, representing poorly managed or naturally growing-rainfall limited mixed vegetation classes (Fig 5-3c).

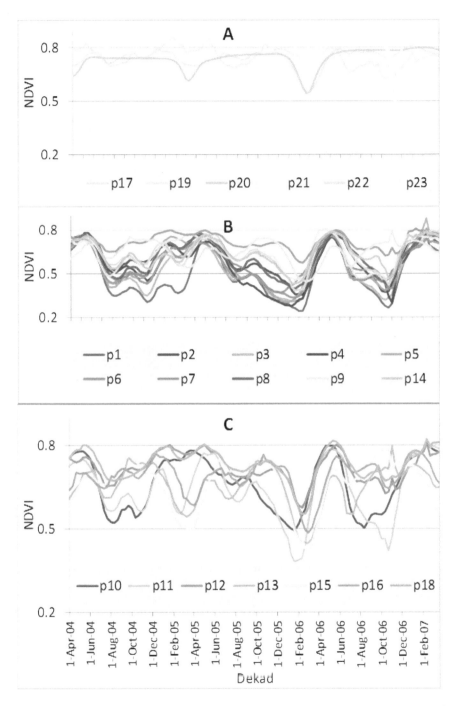

Figure 5-3: Phenological profiles from the ISODATA clusters featuring profiles with no distinct cycles (A), with clearly defined cycles (B) and mixed cycles (C)

The growing season in the Mara basin is rainfall dependent and the basin experiences a high temporal variability of the rainfall (Melesse et al., 2008). It is therefore more plausible to use a cropping calendar in close consultation with the actual seasonality data extracted from the time series for a particular spot within the class. Using this approach, several classes with similar profiles were merged together to form clusters (Fig. 5-4).

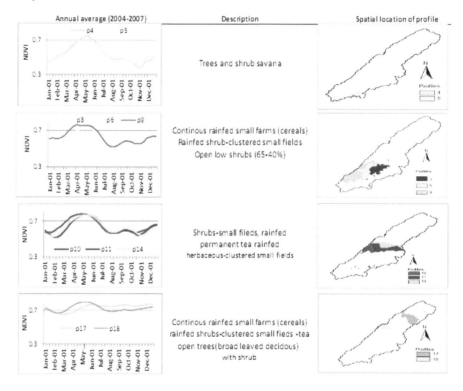

Figure 5-4: Average annual profiles of combined clustered land use classes and their spatial location

The clusters were defined in accordance with the FAO land classification system (LCCS) (Di Gregorio and Jansen, 2000), which is based on classifiers. According to Di Gregorio and Jansen, 1997 classifiers are "a set of pre-selected independent diagnostic attributes defining any land cover class, regardless of its type and geographic location"

Based on local expert knowledge and other available data, notably maps from Google Earth®, the classifiers of the three major land use types present in that cluster were identified. Based on these clusters, a land cover map featuring the main crop/land use types was developed.

The changes in landcover captured by the Landsat images allows for comparison of a specific site to identify changes that "occur slowly and subtly, or quickly and devastatingly"(USGS, 2012). In the humid region, high rainfall amounts lead to healthy vegetation and hence to a saturation of NDVI values which minimise the differences between the different land cover types. Based on local expertise and other available databases, notably maps from Google earth®, the classifiers were identified up to the three major land use types present in that cluster. Based on these clusters, a land cover map featuring the main crop/land use types was developed.

A threshold rule was used to identify the areas with staple crop (maize) and separate them from the other crops. Maize growing in the humid zone (class p15) takes longer to mature compared to that growing in the sub-humid and semi-arid zones (class p6) (Fig. 5-5). The two Landsat images for 2006, representing the green off period in February and green-on period in October were converted to NDVI values and a simple algebraic transformation was used to detect the image difference between them. The image differencing for change detection algorithm which takes two sequence images as input and generate a binary image that identifies changed regions was used.

The separation algorithm subtracts the first date image from the second date image, pixel by pixel. Any class, where the vegetation type was either already growing or not growing at all during the period of the two images will show no change in the differencing map, only vegetation that was not growing at date 1 and still growing at date 2 will show in the binary map. The optimum threshold for change detection of 35% was chosen as the point of intersection for the two cyclic curves. The landuse map at crop level which was driven by the land cover types is given in Fig. 5-6. In order to allow for transferability and consistency, the FAO AFRICOVER legend (Di Gregorio and Jansen, 2000) for land use class naming was adopted.

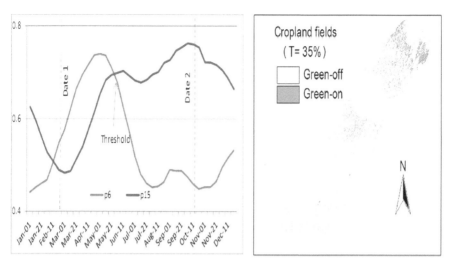

Figure 5-5: Identification of the same crop under different climate zones with phenological profiles (left), and 2-date Landsat image differencing map (right).

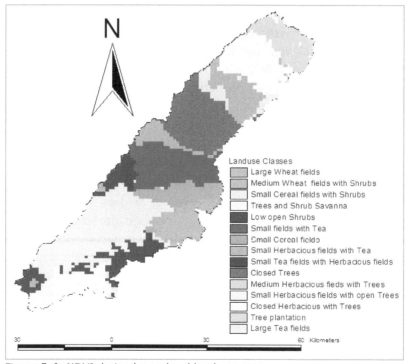

Figure 5-6: NDVI derived crop level land use map

5.4 Accuracy of classification

The quality of thematic maps derived from remotely sensed data should be assessed and expressed in a meaningful way. This is important not only in providing a guide to the quality of a map and its fitness for a particular purpose, but also in understanding the error and its likely implications, especially if allowed to propagate through analyses linking the map to other datasets (Foody, 2002). In thematic mapping from remotely sensed data, classification accuracy is typically taken to average the degree to which the derived image classification agrees with reality or conforms to the 'truth', while, a classification error is some discrepancy between the situation depicted on the thematic map and the reality. Various sources of error including sensor issues, geometric registration, errors introduced by the classification process, assumptions made in the accuracy assessment, and limitations in the map output, accumulate from the beginning of a mapping project through to the end (Lunetta et al., 1991, Congalton, 2005).

The accuracy of the classified maps was determined by use of the error matrix (Congalton, 1991) and by the Kappa coefficient (Congalton and Green, 1993). The kappa value indicates how accurate the classification output is after this chance, or random portion has been accounted for.

$$\hat{K} = \frac{N \sum_{i=1}^{r} x_{ii} - \sum_{i=1}^{r}(x_{i+}*x_{+1})}{N^2 - \sum_{i=1}^{r}(x_{i+}*x_{+1})} \text{---} 5.1$$

where r is the number of rows in the matrix, x_{ii} is the number of observations in row i and column i, x_{i+} and x_{+i} are the marginal totals of row i and column i, respectively, and N is the total number of observations.

Different ground truthing information was used in the accuracy assessment of the classified thematic maps. These included:

- The survey of Kenya's topographic sheets of 1978 was used to assess the 1976 and 1986 maps (Annex 3a).
- The Kenya - Spatially Aggregated Multipurpose Landcover database (FAO Africover, 1999) was used to assess the 1995 map; and
- Field transects conducted in 2010, with secondary data from local offices, were used to assess the 2000 map (Annex 3b).
- The FAO cover map with the LCCS clusters was used for the NDVI map

The LCCS method was used to develop the FAO Africover map. LCCS is a hierarchical, "a priori"" method, where at each level the defined classes are mutually exclusive. At the higher levels of the classification system few diagnostic criteria are used, whereas at the lower levels the number of diagnostic criteria increases. Criteria used at one level of the classification should not be repeated at another. The lowest hierarchy representing the land use was used to assess the NDVI map, while higher hierarchy was used for the landcover Landsat maps. The FAO Africover map was used as a reference map in the error matrix, to assess the image classification accuracy of the NDVI based land use map, while ground truth data was used on the Landsat based land cover maps. The developed NDVI based land use map had an overall accuracy of 83% and a Kappa of 0.79, while the Landsat based land cover map had an overall accuracy of 81% and a Kappa of 0.76. The NDVI map represents larger sampling units leading higher accuracy of the land cover map. This may be attributed to the lumping of land use classes.

Land cover maps are more generic and accommodate a wide scope of land uses under the same land cover. Also, the classes as proposed by the Africover map do not represent pure classes but are in reality a mix of land cover types (Wischut, 2010). According to Kiage et al. (2007) in the absence of aerial photographs, familiarity with the study area and topographic maps proved very helpful for assessing the accuracy of the classification. This line of thought was used for the 1976 and 1986 cover maps, although afew changes might have taken place especially by the 1986.

5.5 Creation of Leaf Area Index (LAI) maps

The LAI can be used for detection of change and for providing information on shifting trends or trajectories in land use and cover change. LAI could be used to validate canopy photosynthesizes models which simulate growth and canopy development based on climate and environmental factors. It is also a sensitive parameter for the control of evapotranspiration in Soil-Vegetation-Atmosphere-Transfer schemes within the context of General Circulation Models (GCMs). The LAI, and ancillary information contained in a single hdf file were downloaded from the VITO webpage. The time series for the LAI were extracted using the in-built smoothening filter in the TIMESAT tool (Jönsson and Eklundh, 2002). Three filters available in the TIMESAT tool where tested for their suitability in mimicking the profile of the LAI time series (Fig 5-7). Compared to the Gaussian and Logistic filters, the Savitsky-Golay filter,

(Savitsky and Golay, 1964) smoothens the time series without substantially changing the amplitude or frequency of the series. The Gaussian and Logistic filters are best suited for extraction of seasonality parameters as they approximated better the coarse seasonality. In order to better reflect the phenology and reduce noise, the Savitsky-Golay filter was used for the LAI filtering in this study. To get the physical LAI values the extracted product's digital number was divided by 30, thus;

$$LAI = \frac{LAI\_DN}{30} \text{--5.2}$$

Where; LAI = physical value (0 to 8.5), LAI_DN= digital number (0 to 255)

The steps involved in the generation of time series profiles for the LAI involved the identification of the land cover type of interest in the thematic map, the overlaying of the classified land use map on the co-registered LAI map and the extraction of LAI information from that point using the TIMESAT tool. Timesat can process data for separate land cover classes. An image file that assigned a code representing a land cover class (0–255) to each pixel needs to be present. By defining an area under a given land use class, all pixels in that land use class were extracted and an ensemble mean calculated. Figure 5-8 shows the process followed in extracting LAI profiles and the resultant time series for the different land use classes.

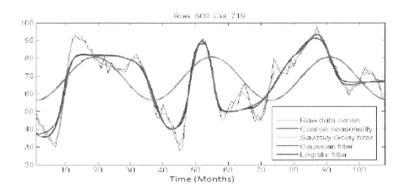

Figure 5-7: The effect of filter algorithms on the characteristic of a time series profile

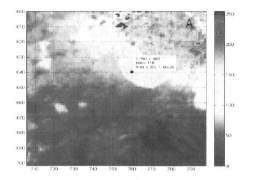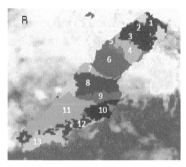

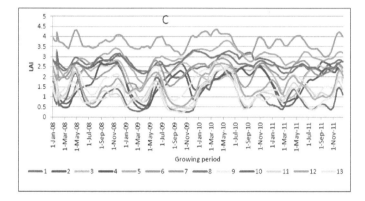

Figure 5-8: The extraction protocol for remotely sensed LAI data A: location , B: overlay of the land use map with classes, 1= tree plantation, 2= upland maize, 3= herbacious crop, 4 = rainfed herbaciuos crop, 5= rainfed tea crop, 6 = Forest, 7 = rainfed shrub crop, 8 = small cereal fields, 9 = small cereal fields with tea, 10 = rainfed wheat fields, 11 = lowland maize, 12 = low open shrubs, 13= trees and shrub savanna, and C: extraction of the different land use classes using TIMESAT tool.

5.6. Conclusion

Different types of public access remotely sensed data were used to create thematic maps for the study area. The landsat 30 x 30m maps were used to generate land cover maps while the SPOT-VGT images were used for the land use maps. LAI maps were also prepared using remotely sensed Leaf area index data. The data output maps were tested for accuracy and the performance assessed as satisfactory.

6. Soil, Crop yield and Water Quality analysis

6.1 Introduction

The Food and Agriculture Organization (FAO) classifies the principal farming systems in Africa either as maize-mixed, cereal-root crop mixed or root crops. These three systems support 41% of the population. Further small holder farmers in Africa maybe categorised based on three criteria: (i) the agro-ecological zones in which they operate; (ii) the type and composition of their farm portfolio and landholding; or (iii) on the basis of annual revenue they generate from farming activities (Dixon et al., 2001). The sizes of the landholding differs depending on the population densities ranging form less than an acres to 10 Ha. It's difficult to estimate crop yields in Africa due to the complex production and land tenure systems. More important is non-uniformity of cultivated plots, failure to harvest all planted areas and enormous post-harvest losses.

In line with integrated water resources management (IWRM), there is an increasing need for the stakeholders to better manage sources of pollution in the river basins. This is especially the case in catchments with changing land use practices and with changing climatic conditions. Excessive loading of organic matter and nutrients into water resources is of a major concern to water resources managers.

6.2 Soil fertility analysis

Soil samples were collected from selected representative farm sites across the basin (Fig. 6-1). Every sample consisted of between 20 - 30 cores taken from the sampling site in a zig zag random sampling pattern (Carter and Gregorich, 2006), air dried and homogenised (Ryan et al., 2001). Atypical areas such as eroded areas, fence lines, roadways, water channels, manure piles, and field edges were avoided during sampling (Carter and Gregorich, 2006). A representative 2 kg composite sample was collected from the homogenized soil. The comparatively inexpensive, composite sampling provides no assessment of field variability and relies on the ability of the farm operator to identify portions of the field that may have inherently different nutrient levels.

The results of the analysis for parameters considered as critical and the optimal ranges for maize cultivation are given in Table 6-1. Results from the soil samples indicated that some soil parameters were outside the FAO (1998) recommended ranges for agricultural crops (maize) production. The soils are also varying from one location to another. Catchment wide the pH, phosphorous, copper and boron levels are below the recommended ranges for maize production. Other elements like calcium, magnesium, sulphur and aluminum were below the recommended range in at least three of the sampled locations. Generally, the soils in the middle section locations (K1, K4) are better suited for maize cultivation than those in the upstream section (K5, K6) and the downstream sections (K2, K3).

6.3 Crop yield studies

6.3.1 Study design for crop yields survey

A stratified multistage cluster design was adopted for the selection of the survey sites with respect to the crop yields in this study (Fig 6-1). This involved a 3-tier selection process for eligible locations to be included in the survey. All the three counties namely Nakuru, Bomet, Narok were included. All the 22 divisions in the study area were also included. By assigning numbers 1 to 55 for all the 55 locations, a random table generator was used to pick 17 completely randomised numbers. There are three reasons for this;

- the use of multistage design controls the cost of data collection,
- the absence or poor quality of listings of households or addresses makes it necessary to first select a sample of geographical units, and then to construct lists of households or addresses only within those selected units.
- The study area exhibits extreme variations in environmental conditions, making it imperative to have samples from all the counties (level 3) and divisions (level 4) to be included.

The study was conducted in the months of July and August 2011. Groundwork and preparation of the study materials, including questionnaires (Annex 4) and soil sampling protocols and gear, was done in July while the actual administering of the questionnaires was undertaken in August. A total of 102 farmers spread over 17 administrative locations in the three counties were interviewed. The design frame involves the selection of at least 5 farmers from each of the 17 pre-selected locations. The choice of farmers depended on the area of the location. Prior to administering the interviews, some ethical issues were addressed.

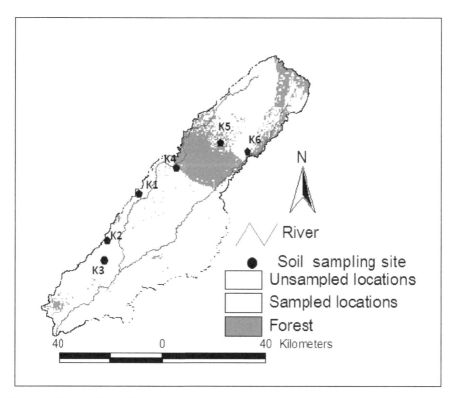

Figure 6-1: Locations of sampled administrative units in the study area

Respondents' anonymity and confidentiality was assured with inclusion of the "informed consent clause" in the questionnaire, the nature of the interview clearly explained and permission requested. Since the interviewer is of a different ethnic group, the responses could have been affected, especially when informants are of low-income status (Schuman and Converse, 1971). Local people were therefore trained to assist in conducting interviews in order to address this issue. This study used the farmer estimation methods (both recall and prediction) as opposed to crop cuts for the determination of farm yields. Estimating crop production through farmer interviews involves asking farmers to estimate for an individual plot, field or farm what quantity they did harvest or what quantity they expect to harvest. As harvest quantities are farmer estimations, they are generally expressed in local harvest units (e.g. sacks, debes,or gorogoros) instead of kg or tonnes. To convert harvest quantities to standard units, conversion factors are required (Table 6-2). To estimate crop yield, production data obtained from farmer recall requires division by the plot area from which the crop was harvested. In order to calibrate farmers' estimation methods, the calibration of their instruments of weighing was done using calibrated weighing scales.

Table 6-1: Results of soil analysis in selected sites in the Upper Mara basin.

parameter	units	Recommended range (FAO, 1998)	Soil Sampling sites					
			Midsection		Downstream section		Upstream section	
			K1	K4	K2	K3	K5	K6
pH		6.00 - 6.80	5.8	5.85	5.81	6.88	4.99	5.7
Phosphorous	ppm	30 - 100	3	2	13	10	2	11
Potassium	ppm	246 - 656	485	503	542	596	341	1565
Calcium	ppm	2524 - 2944	2205	2189	5785	4648	803	864
Magnesium	ppm	252 - 404	355	278	410	257	120	414
Manganese	ppm	100 - 300	207	190	147	198	164	50
Sulphur	ppm	20 - 200	11	11	15	14	16	32
Copper	ppm	2.00 - 10.00	0.59	0.56	1.31	1.01	0.39	0.52
Boron	ppm	0.80 - 2.00	0.6	0.46	0.6	0.61	0.59	0.5
Zinc	ppm	4.00 - 20.00	19.32	6.02	3.97	3.3	20.41	9.22
Sodium	ppm	<242	34	40	112	124	30	55
Iron	ppm	80 - 300	137	136	125	85	169	99
CEC	meq/100g	15.00 - 30.00	21.03	19.66	46.58	29.28	12.72	17.15
Aluminium	ppm	<1200	1219	1038	605	500	1402	1454
EC	us/cm	<800	59	133	126	362	66	200
Org.matter	%	2.50 - 8.00	6.86	4.01	7.28	3.75	6.34	8.45
Nitrogen			-	0.19	0.26	0.12	0.22	0.41

Table 6-2: Conversion of local units into international standard units.

	Commonly used units	SI unit	SI equivalent
1.	Acres	Hectares	
2.	Potato bag	Kgs	110kgs
3.	Maize Bag	Kgs	90kgs
4.	Debe	Kgs	15kgs
5.	Gorogoro tins	Kgs	2.5kgs

6.3.2 Production systems

The estimation of crop yield thus involves both estimation of the crop area and estimation of the quantity of product obtained from that area:

$$\text{Crop Yield (Tons/ha)} = \left[\frac{\text{Harvested crop (tons)}}{\text{Cropped Area(Ha)}}\right] \text{-----------------------6-1}$$

The study found that different food crops are grown in the various locations. Maize was the most commonly grown crop, being cultivated across all locations. Vegetables are grown in the higher altitudes with high precipitation, upper and mid-section areas grow sweet potatoes and wheat is the common complementary crop in the lower semi-arid areas. Some areas have more than four different crops whereas in others only two crop types are grown. The extent in diversification in production systems for the different locations and the average yields for the various crops grown in the locations are given in.Fig. 6-2. The spatial variation in the different crop yields are given inFigure 6-3. The diversification of cropping is a function of the rainfall pattern. In the upper sections, where there is rainfall almost year round, there are more crops grown than in the semi-arid regions. Also, the cropping patterns in the Bomet District are closely intertwined with the rainfall patterns. During the long season (November – May) almost 100% of the farm families perform cropping as compared to 50% - 60% of farm families who undertake cropping during the short season (June – October).

6.3.3 Maize production

Maize growing was carried out in all the selected households. The mean yields varied widely amongst the farmers within a location and amongst the locations. The productivity was the highest in the plain areas of Olulunga and Melelo. Coincidentally these areas neighbour the large scale commercial farms and the average family land holding is higher. The use of mechanized production systems, including plowing with tractors, harvesting with combined harvesters, and application of fertilizers are practised.

The variability in these high producing areas is also very high, indicating differences in the management practices of the farmers. The study area average maize production of 2.65 mton/ha (SD 1.08) is above the national average of 1.6 mton/ha (GOK, 2007) and slightly below the world average (Salami et al., 2010) (Fig. 6-4). Only four administrative locations, all located in the semi-arid regions, had maize yields lower than the national average (Nyoro, 2002).

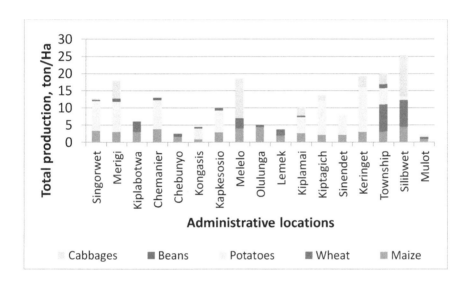

Figure 6-2: Crop diversification in the upper Mara catchment

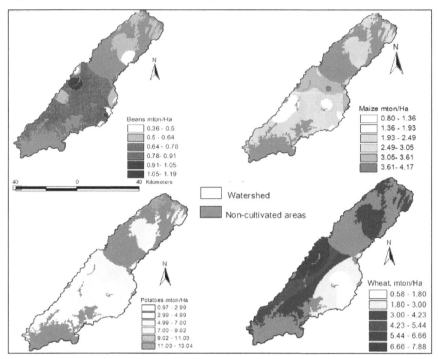

Figure 6-3: Spatial distribution of crop yields for the main crops grown in the study area

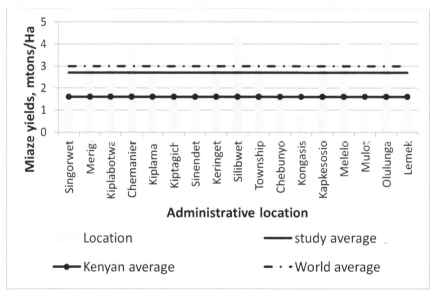

Figure 6-4: Comparison of average maize production in the study with national and global averages

6.3.4 Fertilizer use

The results of the survey show that the use of fertilizer in the study area varies widely. In the highland areas of the upper and midsections of the catchment, there is important use of fertilizers, especially where there is both the intensive production of vegetables and the growing of tea. Farmers having experienced the benefits of using fertilizers on their tea farms embraced the idea of using fertilizers in food production. Tea factories usually supply farmers with fertilizers at subsidized prices or on credit (DAO, verbal communication). The use of fertilizers is also common in the relatively large scale commercial maize and wheat growing areas. According to NEMA (2009) the average level of the use of fertilizers in the Bomet district is low due to the economics and low levels of awareness with regard to the optimum fertilizer levels required for various types of soils. Major fertilizers used are super ammonium phosphates, di-ammonium phosphate (DAP: 20-20-00 and 18-46-00) mainly during the maize planting season of February–March, and calcium ammonium nitrate (CAN) and urea for top-dressing in maize cultivation.

6.4 Water quality investigations

6.4.1 The study design for water quality

Sampling sites were selected in the downstream section, based on the level of human interference (low vs high human impact) and on the availability of historical records from previous water quality studies, to allow for a comparative evaluation of long term temporal change. Bi-weekly monitoring was done in the months of February, March and April, 2011 for 6 primary sites while grab monitoring was carried out in secondary sites spread out in the study area (Fig. 6-5). The primary sites included three sites each from the Nyangores River (Bomet, Silibwet, Masese) and the Amala river (Mulot, Kapkimolwa and Matecha) tributaries.

Sixteen (16) out of the sampled 36 sites, labelled K1-K18 and described in annex (Labels K4 & K5 were erroneously skipped during initial coding) were used to perform a detailed physico-chemical and biological monitoring. For nutrient monitoring, a total of 22 (Annex 5) sites, including the 16 above, were sampled at least once. In situ determinations were performed for pH (WTW pH330i), electrical conductivity (WTW 314i), dissolved oxygen (WTW Oxical), and turbidity (HACH DR890 colorimeter). Samples for physico-chemical analysis were collected using HDPE bottles and transported in cooler boxes to the laboratory where they were stored in a freezer at 18°C, without adding chemical preservatives, until analysis.

Physico-chemical parameters and major ions, including pH, alkalinity, total dissolved solids, free carbon dioxide, total hardness, calcium, sodium, magnesium, potassium, sulphates, carbonates, bicarbonates and chloride, were determined following standard methods of analysis (APHA-AWWA-WPCF, 1985). The following standard procedures were used for the physical-chemical determinations using volumetric (titrimetric), colorimetric, and instrument methods.

Free CO2: Titrimetric method - 50 ml are titrated using 0.0227M NaOH and phenolphthalein indicator.

$$\text{mg CO2/l} = \text{titer x 20 x dilution factor} \text{-----------------------------------} 6\text{-}2$$

Nitrates: 1ml of H_2SO_4 amide was added to 50 ml sample and a blank and let to stand for 7 minutes. 1 ml NED (N-(1-Naphthyl) ethylene diamine dihydrochloride) was then added to the sample and blank and let to stand for another 5 minutes. The colour change to pink was determined using a photometer.

Fluoride: Ion-selective electrode method

Standards of 1, 5 and 10 ppm were prepared from stock solution of 100 ppm. 25ml of each was taken and 25 ml of Tisab buffer added. The fluoride meter was standardized. 25ml of buffer was added to 25ml of the sample and the readings read off the meter. The standard readings were used to plot a graph which was used to determine the fluoride concentration.

20 min Pmv (Dissolved oxygen)

10 ml each of sample and blank were taken and 0.5ml of 4M H_2SO_4 and 2ml of 0.002M $KMnO_4$ were added to both. The sample and blank were then transferred to a water bath and boiled for 20 minutes, and cooled to room temperature. 1ml KI was added and then titrated against 0.01M Na2S2O3 with starch indicator changing colour from blue to colourless.

$$\text{mgO2/l} = \text{blank} - \text{titer x 31.6 x 0.25} = \text{blank} - \text{titer x 7.9} \text{-------------------} 6\text{-}3$$

Turbidity

Standards of 5, 10, and 20 ppm were prepared and the meter standardized. Sample was put into a cuvet and the reading taken from the meter.

Sulphate: Turbidimetric method

Standards of 5, 10, and 20 ppm were prepared and the meter standardized. 2.5 ml of sulphate buffer were added to 50 ml sample, then a spatula full of barium chloride. The sample was stirred for 1 minute transferred to a cuvet and reading taken(X).

$$\text{mg SO4/l} = \frac{X-Y}{3.5} \text{---6-4}$$

Total Iron: Phenanthroline method

2ml conc. HCl and 1ml Hydroxyl amide ($NH_2OH.HCl$) were added each to 50 ml of sample and blank and shaken. It was boiled to 20 ml, let to cool, and then transferred to 50 ml volumetric flask. 10 ml Ammonium acetate ($NH_4C_2H_3O_2$) buffer and 4 ml phenanthroline solution were added and let to stand for 10 minutes. Concentration read off at 540 nm

$$\text{mg Fe/l} = \text{meter reading x 3.33 ----------------------------------6-5}$$

Chloride: Argentometric method

2 drops of Dichromate chloride indicator were added to 50 ml sample and blank, which was titrated against 0.014N $AgNO_3$

$$\text{mg Cl/l} = (\text{titer} - 0.5) \text{ x } 10 \text{ --6-6}$$

pH: Electrometric method

The pH meter was calibrated using buffers of 4, 7 and 9. The pH electrode was inserted into 50 ml sample, and the pH read-off after meter stabilized

Alkalinity: titration method

50 ml of the sample was measured into a flask, a magnet rod dropped in, and the flask placed on a magnetic stirrer. 0.02N H_2SO_4 acid was titrated to pH 4.5, when there was colour change.

$$\text{mgCaCO3/l} = \frac{A \text{ x N x5000}}{\text{Vol.sample}} \text{---6-7}$$

Where: A = ml standard acid used , and N = Normality of Acid

Total hardness: EDTA titrimetric method

1ml total hardness buffer and half full spatula of Eriochrome black T dye were added to 50 ml of sample, and titrated to blue colour using EDTA (ethylenediaminetetraacetic acid).

Calcium: EDTA titrimetric method

1ml NaOH and half full spatula of murexide (Ammonium purpurate) indicator were added to 50 ml sample, and titrated to purple endpoint.

$$mg\,Ca/l = titer \times 8 \text{ --}6\text{-}8$$

Sodium and Potassium: Flame atomic absorption spectrometric method

Sodium standard were nebulized such that 3 ppm read 60, 5 ppm read 100 and 2ppm read 40. The filter was used at 589 nm.

$$mgNa/l = \frac{photometer\ reading}{20\ x\ dilution\ factor}\text{-------------------------------------}6\text{-}9$$

Potassium standard were nebulized such that; 5 ppm read 50 and 2 ppm read 20. The samples were then nebulized

$$mgK/l = \frac{photometer\ reading}{10\ x\ dilution\ factor}\text{----------------------------------}6\text{-}10$$

Manganese: persulphate method

5ml of mixed reagents were added to 100 ml of sample and added to 90 ml. Ammonium persulphate was added, boiled for 1 minute and cooled in running tap for 1minute. It was then transferred to digestive tubes, topped to the mark and transferred again to standard tubes. The colour was compared to standard charts.

 Electrical conductivity and TDS

These were determined using the conductivity meter; the electrode was inserted into the sample. The conductivity was read off first, a selector knob was used to switch over and the TDS read off.

To minimize the error due to the removal and transformations, total phosphorus was determined by the ascorbic acid method on an unfiltered samples. The advantages of the ascorbic acid method is that it produces colour development which is more stable and the reaction using ascorbic acid is independent of temperature and salt concentrations (Jarviel et al., 2002).

The determination of the total nitrogen was based on the persulfate digestion method. Digestion with persulfate oxidizes all forms of nitrogen to nitrate. Nitrate is reduced to nitrite when passed through a copperised cadmium reduction column.

Figure 6-5: Locations of the water quality sampling sites for the different tributaries

The build-up of suspended matter in the reduction column restricts sample flow. The column was flushed regularly to clear the build up of the suspended material. Sample turbidity may also cause interference. Suspended matter was removed by filtration (thus reducing turbidity) of the digested solution through a 0.45 μm pore diameter membrane filter prior to analysis. The persulfate digestion method produces low toxicity waste and is less cumbersome than the classical total Kjeldahl Nitrogen (TKN) methods, but is still sensitive and reliable for extremely unproductive lake and stream samples (Ameel et al., 1993). The detection limit for the analysis of TN and TP was 0.01 mg N/L and 0.001 mg P/L respectively.

6.4.2 Hydro-geochemical classification

The Gibbs' diagram method (Gibbs, 1970) was used to determine the major natural mechanisms controlling the water chemistry. The diagram shows the weight ratio $Na^+/Na^+ + Ca^{2+}$ on the x-axis and the total salinity on the y-axis. According to the diagram, the general environmental origin of the chemicals of the river water in the Upper Mara is mostly (81%) of rock weathering dominance (Fig.6-6). All sites in the Mara mainstream are rock weathering, while 86% and 71% of the sites in the Amala and Nyangores respectively have a rock weathering dominance. The remaining sites are in the atmospheric precipitation dominance zone.

The major cations that characterize the end-members of the world's surface waters are Ca^{2+} for freshwater bodies and Na^+ for high-saline water bodies. The sites are more skewed towards the precipitation than the evaporation dominance. This implies that the study area lies in a recharge area with light mineralization and strong geological influence on the water chemistry. When classified in accordance with the Trilinear diagram method (Piper, 1944), the river water is of the sodium/bicarbonate type, with Na^+ as the dominant cation, while HCO_3^- is the dominant anion. The hydrochemical characteristic of the study area is attributed to the (hydro) geological formation of the area, the rocks in the study area form a wider East African alkaline suite where strongly alkaline series are recognized and characterized by a dominance of sodium over potassium (Saggerson, 1991). The statistical parameters of the measured variables are given in Table 6-3.

At average concentrations, most of the parameters are within the World Health Organization (WHO, 2008) guideline levels for drinking water. Of the parameters measured, only some sites (30%) had levels of iron and manganese even higher than the maximum allowable values. The high manganese levels in the Amala and Mara sub-basins are due to natural sources compounded by low oxidation conditions, and not to anthropogenic contamination, since the area has no known industrial and mining activities.

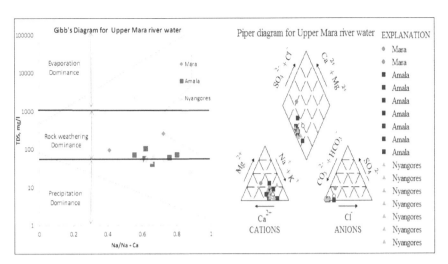

Figure 6-6: Geochemical classification of the river water according to Gibb's (left) and Piper (right) diagrams

The high standard deviation of most variables indicates a strong (spatial) variability in chemical composition among the samples. There is a clear upstream-downstream trend in the pH and electrical conductivity for both the Nyangores and the Amala tributaries. The pH decreases, while the EC increases in the downstream direction for both rivers. The Amala River has higher values than Nyangores for both parameters. The EC for the wet season is lower than during the dry season due to the dilution effect. The pH and EC are consistently changing downstream, even where there is a change in the land use type, indicating that they are less sensitive to the land management process and driven more by the natural conditions. The changes in geological conditions, since the rock composition determines the chemistry of the soil in the watershed and ultimately the river water, are the key determinants of the hydrochemistry in the study area.

6.4.3 Spatial variation of the nutrients (TN and TP)

The variation of nutrient loading in the river water is driven by both natural and human activities. The total nitrogen concentration in samples downstream of the urban centers of Bomet, Silibwet, Longisa and Mulot was higher (>1mg/l) than the rest of the sampled sites. Tenwek hospital, whose wastewater treatment plant is the only known source of point pollution, is located in this region. This indicates that there is potential risk of nitrogen from the numerous onsite waste disposal systems, including pit latrines (Fig. 6-7). The diversification of the agriculture practices in the area, involving the intensive use of fertilizers on both cash crops (tea) and food crops (carrots, cabbages, maize, onions) (§6.3) is also a potential source of high nitrogen concentrations. The low nitrogen loading on the Amala and Lower Mara subbasins may be attributed to deforestation and sparse population density. The spatial distribution of the total phosphorus (TP) is biased to the type of agricultural activities. There is a high TP concentration at the confluence of the Nyangores and Amala rivers and also in the Upper Amala regions. These are areas with large-scale mechanised farming. At the confluence, there are large scale commercial farms under irrigation (Hoffman, 2007). The Upper Amala regions are also characterized by large wheat/barley cultivation where large quantities of commercial fertilizers are potentially used (Fig. 6-7). Phosphorus is usually the limiting nutrient for eutrophication. Soil analysis in the study area have shown that all the sites that were sampled in the mixed farming zones are deficient in phosphates (see §6.2)

Table 6-3: Mean (±SD) physico-chemical water quality parameters for the major sites sampled on the Mara, and its Nyangores and Amala tributaries.

Sampling sites

Parameter	K1	K2	K3	K6	K7	K8	K9	K10	K11	K12	K13	K14	K15	K16	K17	K18
pH	8.31	7.78	7.67	7.8±0.33	7.52	7.5±0.4	7.8±0.37	7.4±0.21	7.49	7.97	7.5±0.3	7.4±0.18	7.64	7.46	7.32	7.81
EC (µS/cm)	±17	114.8	86.7	78.7	84.8	54±4.6	87±11.6	74±6.8	42.4	166.7	154±29	66±13	154	131±30	105.3	96.4
TN (mg/l)	0.65±0.81	0.48±0.15	0.32	1.1±1.3	0.08	0.29±0.32	0.46±0.33	0.70±0.53			0.6±0.58	0.64±0.23	0.06±0.05	0.51±0.26		0.28±0.23
TP (mg/l)	1.45±1.56	0.56±0.76	0.03	0.53±0.52	0.21±0.03	0.28±0.50	0.23±0.39	0.37±0.42	0.13		0.42±0.6	0.26±0.37	0.28±0.11	0.31±0.36		0.32±0.24
Colour	150	<5	<5	<5	<5	<5	10	10	<5	<5	150	<5	50	1050	<5	<5
Turbidity (NTU)	104	6	9	53	12	36	51	63		14	1435	41	1690	1970	5	10
Pmv mgO2/l	79	1.9	1.58	2.7	1.9	2.7	3.1	9.1	1.9	1.9	91	3.9	94	9.1	1.58	2.7
Fe (mg/l)	0.2	0.26	0.21	0.64	0.78	0.54	1.6	2.6	0.8	0.71	16	2.1	18.3	15.9	0.63	0.6
Mn (mg/l)	0.2	0.06	<0.01	0.06	0.07	0.64	0.18	0.32	0.24	1	1.8	0.28	1.2	1.6	0.08	0.24
Ca (mg/l)	22.4	4.8	6.4	5.6	4.8	6.4	7.2	7.2	0.8	11.2	3.2	6.4	12.8	9.6	6.4	4
Mg (mg/l)	5.4	0.49	0.98	0.49	0.49	0.98	0.49	0.49	1.46	3.4	2.43	1.95	5.84	1.46	1.95	2.43
Na (mg/l)	59	19.5	10.4	10.5	12.8	10.5	11.52	12	5.8	18.4	6.2	9.4	8.9	12	12.7	12.6
K (mg/l)	1.4	0.8	0.6	0.6	0.4	0.8	0.6	0.6	0.4	1	0.6	0.6	1.1	1	0.8	0.6
Hardness (mgCaCO3/l)	78	14	20	16	14	20	20	20	8	42	18	24	56	30	24	20
Alk	122	46	36	38	30	36	32	40	20	72	40	34	12	42	42	42
Cl⁻ (mg/l)	6	4	4	1	3	3	4	4	0	7	11	3	14	6	4	3
F (mg/l)	1.4	0.38	0.3	0.31	0.5	0.42	0.28	0.33	0.28	0.35	0.98	0.3	1	1.3	0.38	0.24
NO3⁻ (mg/l)	2	0.98	0.21	0.35	1.8	1.3	2.3	0.21	0.21	0.18	2	1.8	2.7	0.98	0.8	0.36
SO4²⁻ (mg/l)	23.4	0.86	1.43	<0.3	1.7	<0.3	<0.3	<0.3	0.86	<0.3	<0.3	<0.3	31.4	<3.0	2.28	<0.3
TDS (mg/l)	259	71	54	49	53	54	57	58	26	103	40	56	96	71	65	60

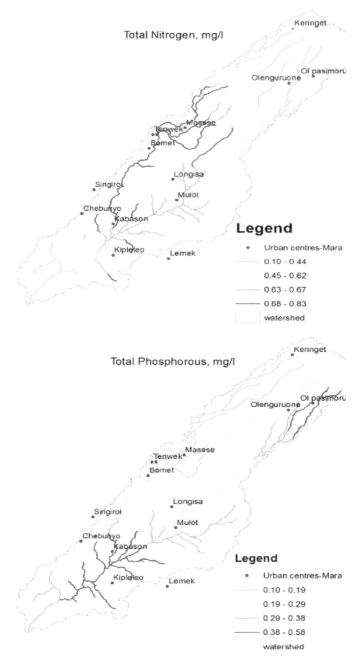

Figure 6-7: Spatial distribution of Total Nitrogen (above) and Total Phosphorus (below) loading on rivers in the Upper Mara basin

6.4.4 Temporal variation of the water quality

There is seasonal (rainy/dry) variation in the physico-chemical quality of the river. The electrical conductivity was generally lower in the wet rainy season for the primary sites due to the effects of dilution. The nutrient concentration levels are very precipitation sensitive with the lowest records corresponding with the dry days and the highest with the wet days. The total nitrogen (TN) and total phosphorus (TP) concentrations ranged from 0.04 mg/l to 2.02 mg/l, and 0.01 to 3.4 mg/l respectively. The level of variability over time and space is best represented by comparing the different sites downstream by use of a box plot (95% confidence interval) (Fig. 6-8). The sites of Masese and Matecha, located close to the forested area, have the lowest temporal variation, while the stations located downstream in the agricultural and densely populated areas have not only the highest variability ranges but also the higher recorded concentrations for both TP and TN. Over the study period, the TN was more variable than TP for all the sites investigated. In the forested areas, there are stable and small fluctuations in the nutrient level.

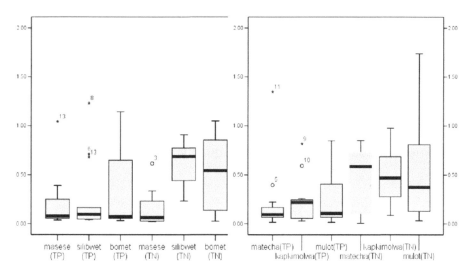

Figure 6-8: Spatial (downstream from right) and temporal (box-height) variation in total phosphorus (TP) and total nitrogen (TN) for three key sites on the Nyangores (left) and Amala (right) tributaries

The high variability of TN at Matecha -located in the forested area- is caused by the presence of large amounts of animals that drink directly in the river, as well as by irrigated vegetable farms in the vicinity. These

73

man-made activities may have contributed to the increased pulses in TP concentrations. The effects of the rainfall on the nutrient concentration were assessed at the Masese monitoring station over a period spanning both the dry and the wet seasons. Both nutrients (TP and TN) show the highest concentrations at the onset of the rainy season in March and April. In February, the driest month, the nutrient concentrations are low.

The highest determined TP levels, on 24th of March 2011, happened one to two days after a storm during which more than 40mm of rainfall fell down within 24 hrs. Similar events were experienced at Bomet on and around the 21st of April 2011. The inability to exactly capture the peak nutrient fluxes is due to the manual sampling equipment used. Since limitations in the knowledge of temporal lags in chemical transport and their causes create uncertainties in the periods over which steady-state conditions apply (Schwarz et al., 2006), an automatic sampling protocol should be used to reduce these uncertainties.

6.5 Conclusion

The soil in the Upper Mara basin are generally of good quality, although they are deficient in phosphorous, a key limiting nutrient. Crop yields in the basin are above the national levels. There is considerable use of fertilizers especially in the upper reaches of the basin. Although the concentration for both total phosphorous and total nitrogen remain low (< 1 mg/l), the concentrations during the wet season are beyond the levels for natural systems, suggesting the influence of anthropogenic interference in agricultural streams. At the current levels of the nutrient concentrations and compared with similar landcover types especially in the developed countries, the adverse impacts of land-use and management practices on the water quality status of the river are minimal.

7. Trend Analysis for Assessment of Climate Variability

7.1 Introduction

According to Osima et al., "In order to detect climate change at a place, rigorous statistical analysis and tests should be performed on climatological variables. Such analysis should include: trends, long term mean change in a climatic variables, changes in frequency and severity of extreme events, temporal distribution of climatic events like rainfall onset and cessation dates, including shifts in seasons".

A trend is a significant change over time exhibited by a random variable, detectable by statistical parametric and non-parametric procedures, (Longobardi and Villani, 2010). According to Sonali and Kumar (2013), there are two common statistical approaches for trend detection in climatic variables. The slope based tests, including least squares linear regression and Sen's robust slope estimator, need to satisfy both distributional and independent assumptions. Rank-based tests, including Mann–Kendall and Spearman rank correlation are nonparametric and need to satisfy independent assumptions only. In case of violations of the assumptions of independence of observations, the serial correlation can be removed by pre-whitening the series or pruning the data set to form a subset of the observations that are sufficiently separated temporally to reduce the autocorrelation (Burn and Elnuur, 2002). Statistical approaches which consider the effect of serial correlation include: the pre-whitening, trend-free pre-whitening, variance correction approaches (Hamed and Rao, 1998), modified Mann-Kendall (Yue and Wang, 2004), and block resampling techniques.

For trend analysis in the study, six rainfall gauging stations situated within the study area and which have longer term data (>30 yrs) were used. Since no temperature measurements were available within the basin, three meteorological stations which sandwich the study area between them were used for the temperature trends analysis. For the vegetative trends analysis, NDVI data from the SPOT-VEGETATION sensor was used. A summary of the trends analysis details is given in Table 7-1. While the locations of the stations and the characteristic of the data used are given in Fig 7-1.

Table 7-1: Variables and methods were used for the trend analysis.

	Variable	Data range	Method(s) used
	Climatic variables		
1	Rainfall	1962 - 2008	Mann-Kendall
			Sen's Slope
2	Temperature	1993 - 2009	Mann-Kendall
			Sen's Slope
	Vegetation variables		
3	NDVI	1999 -2010	Seasonal NDVI
			Integrated NDVI
			Vegetation condition index
			Standard condition index
			Vegetation productivity indicator
4	Landcover change Landsat MSS/TM/ETM+	1976 - 2006	Post classification
			Three date NDVI-RGB

7.2 Trend analysis using climatic variables

Kenya experiences a bimodal rainfall distribution with long rains in March, April, May (MAM), and short rains in October, November, December (OND). The major systems that influence the climate include the InterTropical Convergence Zone (ITCZ), Sub Tropical High Pressure systems (STHP), El Niño/Southern Oscillations (ENSO), Monsoon winds, tropical cyclones, the Indian Ocean, Lake Victoria circulation and the regional topography.

Trend analysis for climatic variables was based on the nonparametric Mann-Kendall test for the trend and the nonparametric Sen's method for estimation of the magnitude of the trend. According to Salmi et al. (2002), in the Mann-Kendall test, missing values are allowed and the data do not need to conform to any particular distribution. The Sen's method is not greatly affected by gross data errors or outliers, and can be computed when data are missing. The mann-Kendall test is used specifically to determine the central value or median changes over time (Helsel and Hirsch, 2002), and the statistic Z (or S if sample size, n<10) is given by:

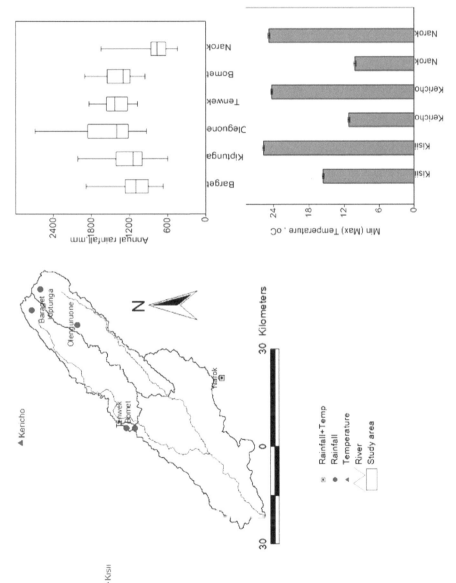

Figure 7-1: Location of the gauging stations and the range in the datasets used in the trend analysis

$$S = \sum_{i=1}^{n-1} \sum_{j=i-1}^{n} sgn(x_j - x_i) \text{-------------------------------7.1}$$

Where:

$$sgn(x_j - x_i) = \begin{cases} 1, & x_i > x_j \\ 0, & x_i = x_j \\ -1, & x_i < x_j \end{cases} \text{----------------------7.2}$$

For a time series x_k, k = 1, 2... n.

When n ≥ 10, S becomes approximately normally distributed with mean = 0 and variance as:

$$\partial_s^2 = \frac{n(n-1)(2n+5) - \sum_t t(t-1)(2t+5)}{18} \text{----------------------7.3}$$

Where: t is the extent (number of x involved) of any given tie and Σ denotes the summation over all ties. Then Z_c follows the standard normal distribution where:

$$Z_c = \begin{cases} \frac{(S-1)}{\partial_s}, & S > 0 \\ 0, & S = 0 \\ \frac{(S+1)}{\partial_s} & S < 0 \end{cases} \text{----------------------7.4}$$

The null hypothesis that there is no trend is rejected when:

$$|Z_c| > Z_{1-\frac{\alpha}{2}} \text{-------------------------------7.5}$$

Where, Z is the standard normal variate and a is the level of significance for the test. At certain probability level H_0 is rejected in favour of H1 if the absolute value of S equals or exceeds a specified value $S_a/2$, where $S_a/2$ is the smallest S which has the probability less than a/2 to appear in case of no trend. A positive (negative) value of S indicates an upward (downward) trend.

The Sen's method uses a linear model for the trend. The magnitude of trend is predicted by the Sen's estimator. The slope (T_i) of all data pairs is computed as

$$T_i = \frac{x_j - x_k}{j - k} \text{-------------------------------7.6}$$

For i = 1, 2... N.

Where: x_j and x_k are considered as data values at time j and k (j>k), correspondingly.

The median of these N values of Ti is represented as Sen's estimator of slope which is given as:

$$Qi = \begin{cases} T_{\frac{N+1}{2}} & N \text{ is odd} \\ \frac{1}{2}\left(T_{\frac{N}{2}} + T_{\frac{N+2}{2}}\right) & N \text{ is even} \end{cases} \quad \text{-----------------------------------7.7}$$

A positive value of β indicates an increasing trend whereas a negative value indicates a decreasing trend.

According to Hamed and Rao (1998) and Yue et al. (2002), trend detection in a series is largely affected by the presence of autocorrelation. Autocorrelation is either positive or negative, and is the correlation of a time series with its own past and future values. A positive autocorrelation is a specific form of "persistence", a tendency for a system to remain in the same state from one observation to the next. With a positive autocorrelation in the series, the possibility for a series to be detected as having a trend is higher; this may not always be true. With a negative autocorrelation, the possibility of a series to be detected is less, hence an existing trend maybe missed.

The graphic method(s) and the Durbin Watson (DW) statistic (Durbin and Watson, 1951) are the commonly applied methods to test for serial correlations in a time series. The graphic methods for assessing the autocorrelation of a time series are: the time series plot, the lagged scatterplot, and the autocorrelation function. The autocorrelation function was carried out on evenly sampled temporal/stratigraphic data. The lag times τ up to n/2, where n is the number of values in the vector, are shown along the x axis, the autocorrelation function is symmetrical around zero (Davis, 1986).

In the DW method, the statistic d provides a test of the null hypothesis H_o: ρ =0 in the following specification for the error terms, $\mu_i = \rho\mu_{i-1} + \varepsilon_t$. If the test is rejected, there is evidence for first-order serial correlation. By checking the DW table for critical values, the above hypothesis can be tested.

$$d = \frac{\sum(u_t - u_{t-1})^2}{\sum u_t^2} \quad \text{--7.8}$$

$$\rho = \frac{\sum(u_t \cdot u_{t-1})}{\sum u_t^2} \quad \text{---7.9}$$

Where: ρ= estimated serial correlation coefficient, and d= 2(1-ρ), If there is no serial correlation, ρ=0, then d=2, If there is positive serial correlation, ρ>0, then d<2, If there is negative serial correlation, ρ<0, then d>2.

In order to test for the serial correction;

Test $H_o: \rho = 0$ against $H_A: \rho > 0$ and $H_o: \rho = 0$ against $H_A: \rho < 0$ for positive and negative serial correlation, respectively. The critical values, d_L and d_u at α of 1%, 5% or 10%, and note k' is the number of coefficients in the regression excluding the constant.

For the positive correlation, the null is rejected if $d \leq d_L$, if $d \geq d_U$, the null is not rejected, if $d_L < d < d_U$, the test is inconclusive, while for negative correlation, the 4-d is computed and the null is rejected if; $4 - d \leq d_L$. If $4 - d \geq d_U$, the null is not rejected, if $d_L < 4 - d < d_U$, the test is inconclusive. The decision zones for the DW test are summarized in Fig 7-4.

7.2.1 Rainfall trend analysis

Trend analysis for rainfall data was performed for 6 stations namely: Baraget, Kiptunga, Olenguruone, Tenwek, Bomet and Narok, which have relatively long term data records. The analysis was done for monthly, seasonal and annual rainfall. Trend analysis was performed for the wet seasons of October, November and December (OND) and March, April and May (MAM), as well as the dry seasons December, JanuaryandFebruary (DJF) and June Julyand August (JJA). The summary of the results of the mann-Kendall trends (Z) and the Sen Estimator (Q) are given in Table 7-2.

Rainfall gauging stations Baraget and Olenguruone in the upper part of the catchment have significant of decreasing trend in rainfall at both the 95 and 99% confidence levels. Olenguruone has significant decreasing trend at three months in a year. The dry seasons of DJF and JJA have a significant decreasing trend in rainfall at 95 and 99% confidence level respectively. The other station experiencing a decreasing trend is the Baraget station. In contrast to Olenguruone the station has a significant trend (95% and 90) in the wet MAM and OND season respectively. Elsewhere, stations in the midsections of the basin namely Bomet and Tenwek have not changed significantly even at the 90% confidence level. In the semiarid regions, the Narok station has had significant decreasing trends in annual rainfall at 90% confidence level. All stations have a decreasing trend for the MAM rainfall season. Only Bomet station has an increasing trend for the DJF, which is however insignificant. The overall annual outlook for the basin is a decline in rainfall with the Sen's estimator for the true slope of the linear trend (change per unit time period) indicating a rate of 3mm/yr to as high as 18mm/yr (Fig 7-2).

The decreasing trend is consistent with the analysis performed by Funk et al. (2010) for the MAMJ period for 1970-2009 over the entire maize surplus region in Kenya, which also include the study area. They linked the decline in rainfall amounts to the warming of the Indian Ocean. Kizza et al. (2009) in a study of trends in the lake Victoria regions found Sotik and Kericho stations located in close proximity (ca. 50km) to the study area to have experience mixed trends over a time from 1925. In the periods 1941-1980, 1961-1990 there was no change in the trends. There was, however, a negative change (decline) in rainfall in the period 1971 – 2005. Mote and Kasser (2007) blamed human activities for being the main cause of deforestation, and relate the decrease in rainfall to deforestation around the Kilimanjaro mountain.

According to Bruijnzeel and Proctor (1995), in tropical montane forests the trees intercept mist and this source of moisture is lost after logging. The "mist harvesting" effect amounts to between 5 and 20 percent of total precipitation. By pumping enormous amounts of atmospheric moisture from the ocean, a forest regulates precipitation to be spatially uniform over the entire catchment; moisture is then returned to the ocean in the liquid state as runoff. Forest climate control by precipitation prevents moisture shortage and droughts, as well as excessive precipitation and floods (Makarieva and Gorshkov, 2010). Water vapour emitted from the trees through evapotranspiration stimulates rainfall whilst the roots reduce the risks of floods and drought by storing water and binding topsoil. Deforestation disrupts this cycle, leading to a reduction in regional rainfall.

The Mau forest areas have behaved in the same manner as analysed by Makarieva and Gorshkov, 2010. There is reduction in the amount of rainfall of upto 18mm/yr (translating to a high of 900mm in 50 yrs). The high drop in the rainfall may be attributed to both the actual decline in precipitation amount or to a lesser extent the quality of available data including missing data. During the generation of annual rainfall, years with more than 15 consecutive days of missing data were skipped.Locals living within the Mau forest area attest to witnessing declining rainfall amounts but their claims could not independently verified.

Table 7-2: Mann-Kendall (Z) and Ser estimator (Q) results for selected rainfall gauging stations (****trend at α = 0.001 level of significance, *** trend at α = 0.01 level of significance, ** trend at α = 0.05 level of significance, and * trend at α = 0.1 level of significance).

Month	Narok		Tenwek		Bomet		Kiptunga		BARGET		Olenguruone	
	Z	Q	Z	Q	Z	Q	Z	Q	Z	Q	Z	Q
Jan	-0.31	-0.19	0.99	0.87	0.84	0.67	0.13	0.04	-0.77	-0.49	-0.72	-0.30
Feb	-0.41	-0.26	0.19	0.35	0.71	0.58	-1.11	-0.37	0.46	0.19	0.04	0.05
Mar	-0.01	0.00	0.26	0.30	-0.48	-0.44	-0.66	-0.33	-0.60	-0.35	-0.48	-0.35
Apr	-0.98	-0.64	-0.41	-0.52	-1.30	-1.47	-1.43	-1.46	-1.05	-1.39	-0.80	-1.17
May	-0.65	-0.44	-1.21	-1.19	1.58	1.60	-0.77	-0.51	-2.2**	-2.21	-2.68***	-3.44
Jun	-1.51	-0.27	1.28	0.66	-1.95*	-0.97	-0.57	-0.32	1.29	0.88	-1.19	-0.62
Jul	-0.22	-0.02	-0.15	-0.12	-1.03	-0.48	0.62	0.38	-0.07	-0.04	-2.27**	-1.73
Aug	0.73	0.14	1.74*	1.05	0.73	0.56	-0.03	-0.03	-1.10	-0.95	-1.96*	-1.91
Sep	-0.78	-0.14	0.11	0.09	0.69	0.31	-1.18	-0.68	-0.83	-0.63	-1.54	-1.46
Oct	-1.10	-0.19	0.53	0.29	-1.27	-0.83	0.19	0.07	-1.08	-0.65	-0.30	-0.25
Nov	-0.54	-0.23	0.76	0.92	1.21	1.44	0.03	0.01	-1.10	-0.74	-0.81	-0.55
Dec	-0.97	-0.51	-0.79	-0.72	0.06	0.09	-0.35	-0.19	-0.40	-0.23	-0.86	-0.50
Seasons												
DJF	-1.13	-1.35	-0.13	-0.33	-0.27	-0.56	-0.30	-0.43	-1.54	-3.00	-2.50**	-4.74
MAM	-0.60	-0.90	-0.73	-1.54	0.27	0.77	-1.59	-2.56	2.47**	-4.73	-1.51	-3.05
JJA	-1.03	-0.33	0.55	0.62	-0.51	-0.76	0.71	1.12	-0.06	-0.14	-3.58****	-8.18
OND	-1.40	-1.21	0.40	0.87	0.74	1.43	-0.30	-0.43	-1.81*	-2.00	-1.39	-2.63
Annual	-1.72*	-3.19	-0.69	-1.96	0.31	0.95	-1.38	-4.85	2.36**	-9.19	-3.06***	-18.34

82

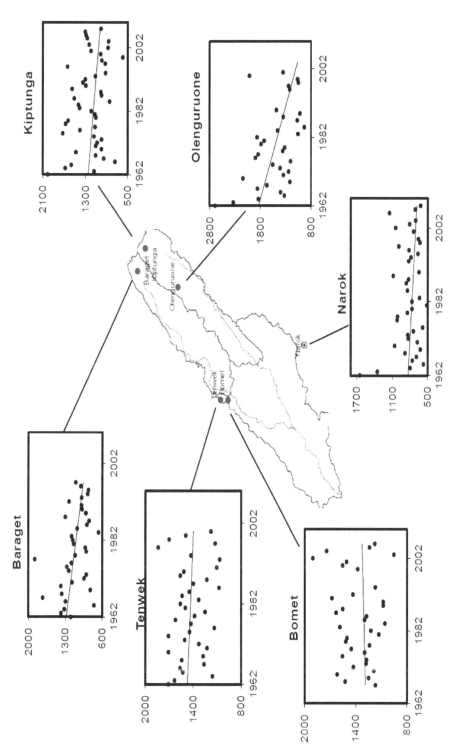

Figure 7-2: The Sen slope (line) for the observed rainfall data in stations within the Upper Mara basin.

7.2.2 Autocorrelation test

The graphical autocorrelation test (Fig. 7-3) indicates that the data from all the stations is symmetrical around zero at a 95% significance. A predominantly zero autocorrelation signifies random data and non serial correlation.

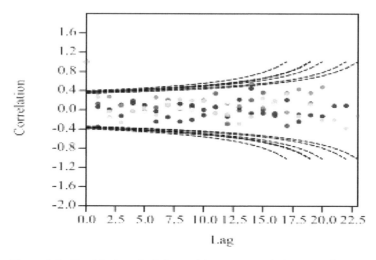

Figure 7-3: Graphical method for positive autocorrelation test of annual rainfall

Positive autocorrelation	Zone of indecision	No Autocorrelation	Zone of indecision	Negative autocorrelation
0 d-lower	d-upper	2 4-d-upper	4-d-lower	4

Figure 7-4: Decision chart for the Durbin Watson statistic analysis of autocorrelation

The results of the autocorrelation as determined by the DW statistics indicate that the rainfall is not serially correlated. The statistic tested the violation of an assumption of Ordinary Least Squares (OLS) regression for residuals in the data. Five of the six stations have d-values very close to the zero autocorrelation value of d=2, as shown in Table 7-3. These five stations have d-values either greater than d_{upper} for those with d<2, or less than 4-d_{upper} for those with d>2. This implies that the stations have time series data which satisfy the condition for a zero autocorrelation at 95% confidence level. For the Bomet station, d>4-d_{Lower}, hence the time series has a clear negative autocorrelation at 95% confidence level. Although the Mann-Kendall trend test indicated that

84

Bomet has a non significant increasing trend even at 10% significance level, this trend may have been missed due to the negative autocorrelation of the data series.

Table 7-3: Autocorrelation analysis with the Durbin Watson statistics.

Station	calculated d- value	Positive autocorrelation d_{Lower}	d_{upper}	Negative autocorrelation 4-d_{upper}	4-d_{Lower}
Barget	2.052			2.481	2.598
Kiptunga	1.839	1.475	1.566		
Olenguruone	1.679	1.442	1.544		
Tenwek	2.288			2.456	2.558
*Bomet	2.679			2.456	2.558
Narok	1.589	1.503	1.585		

7.2.3 Temperature trend analysis

To determine the trend in minimum and maximum temperatures, the average monthly and annual temperatures were established for Kisii, Kericho and Narok stations, which though not located within the study area, are strategically located round the basin, and are the only station with long term temperature data (1992-2009). The trend analysis and the estimated quantity of the change are given in Table 7-4. The Mann-Kendall trend indicates that the annual minimum temperatures have increased significantly with 95% significant level for Kisii and Kericho stations and 99% for the Narok station (Fig 7-5).

The significant increases in the minimum temperatures have been observed between the months of July and November for all the stations. There is no significant change in the maximum temperature over the study period. The Mann-Kendall statistic has a Z~0 for Kisii and Narok indicating there is no direction in the trend. Kericho has a positive increasing trend although insignificant even at 90% significance level. The month of April has experienced a negative Z score in the monthly maximum temperatures for two of the three stations with the decline in Narok being significant at $a = 0.1$

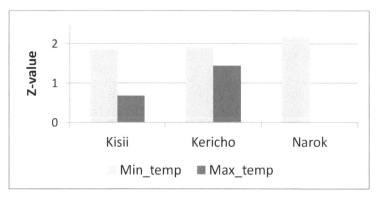

Figure 7-5: Mann – Kendall trend analysis for mean annual minimum and maximum temperatures

The rate of change in minimum temperature ranged between 0.02 to 0.04^{0}C (Fig 7.6), while the maximum temperature changed between 0.01 to 0.02. Hulme et al., 2001 have shown that Africa has experienced an increase in temperature of 0,05°C per decade in the 20th century. While the changing climate may be responsible for the rise in temperature, anthropogenic causes may also be playing a significant role. The increase in temperatures may be partly explained by the increase conversion of landcover type from permanent cover types like forest and shrubs to cultivated fields. According to Pokorný, 2010, "transforming landscapes from forest to field has at least as big an impact on regional climate as greenhouse gas–induced global warming".

Hesslerova and Pokorny, 2010, Pokorný, 2010, compared the surface temperature of tea plantations, rain forest; and farmland. Despite having the highest amount of chlorophyll (being the greenest), the temperature of tea plantations ranges between 30 –35 °C, that is more than in case of forest. The highest temperature is characteristic for the crops (35 –45°C), depending on the crop cover, type, wetness, and other factors. This shows that the surface temperature depends on the type of land cover and confirms forests as the coldest landscape segments.

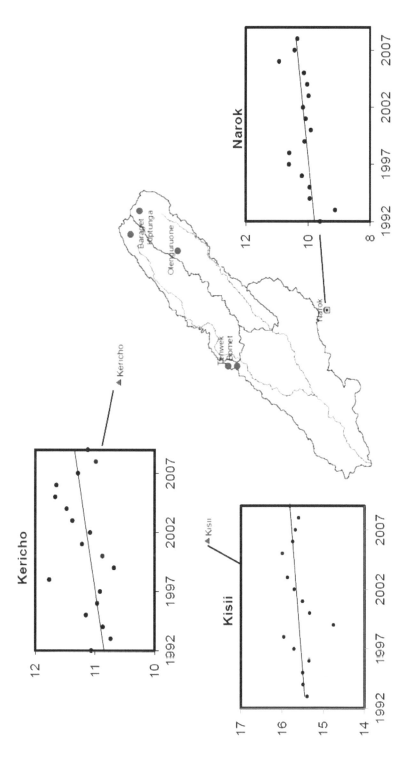

Figure 7-6: The increment in minimum temperature (line) for the stations located around the Upper Mara basin.

7.2.4 Spatial distribution of the climate change

The spatial extend of the trends in the temperature and rainfall was assessed by ploting the quantity of change in the climatic variables given by the Sen's slope in the trend analysis on the basin map. The rainfall decreases outwards from the Olenguruone station which has the highest decrease to Bomet where a slightly positive trend is observed (Fig 7.7). The Colonial Government's setting up of the Olenguruone Settlement Scheme in 1941 for the resettlement of natives displaced by the white settlers in the tea growing zones in Kericho/Tinet area (Kimaiyo, 2004). The land resettlement exercise intensified between 1969-1991 with more conversion of forest to agricultural land (Jama 1991).

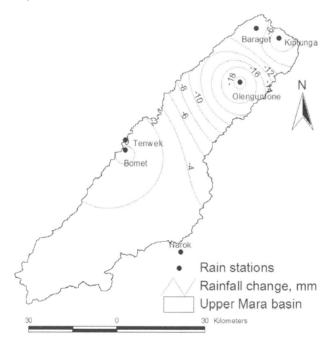

Figure 7-7: Spatial variability of the changes in rainfall for stations within the Upper Mara basin (1962-2008

Table 7-4: Mann-Kendall (Z) and Sen estimator (Q) results for selected temperature gauging stations (*** trend at α = 0.001 level of significance, ** trend at α = 0.01 level of significance, ** trend at α = 0.05 level of significance, and * trend at α = 0.1 level of significance).

Gauging stations

Month	Kisii				Kericho				Narok			
	Z_{min_temp}	Q_{min_temp}	Z_{max_temp}	Q_{max_temp}	Z_{min_temp}	Q_{min_temp}	Z_{max_temp}	Q_{max_temp}	Z_{min_temp}	Q_{min_temp}	Z_{max_temp}	Q_{max_temp}
Jan	1,22	0,037	1,22	0,095	0,68	0,021	-0,37	-0,030	0,62	0,064	-0,29	-0,015
Feb	1,13	0,022	0,68	0,048	0,83	0,053	0,29	0,026	0,50	0,026	0,14	0,023
Mar	0,68	0,015	-0,59	-0,033	0,23	0,015	-0,23	-0,010	1,28	0,076	-0,87	-0,047
Apr	1,40	0,027	0,05	0,005	0,76	0,027	-0,53	-0,022	0,12	0,007	-1,77*	-0,071
May	0,77	0,012	1,31	0,050	1,44	0,036	-0,12	-0,004	0,41	0,021	0,50	0,012
Jun	0,90	0,012	1,08	0,041	-0,19	-0,005	0,45	0,019	-0,70	-0,044	-0,78	-0,021
Jul	3,20***	0,039	-0,05	-0,005	0,68	0,023	2,18**	0,047	0,29	0,015	0,62	0,014
Aug	2,43**	0,047	-0,30	-0,019	1,06	0,021	0,70	0,032	2,43**	0,085	0,37	0,016
Sep	0,95	0,019	0,20	0,010	2,12**	0,058	0,12	0,001	3,17***	0,166	-0,37	-0,012
Oct	2,03**	0,029	0,23	0,041	0,45	0,016	0,23	0,008	2,48**	0,143	-1,29	-0,028
Nov	2,21**	0,030	0,14	0,013	-0,23	-0,003	0,68	0,031	0,37	0,060	1,40	0,060
Dec	0,44	0,020	-0,59	-0,043	0,98	0,025	0,45	0,027	0,30	0,034	0,99	0,099
Annual	1,85*	0,022	0,68	0,012	1,89*	0,029	1,44	0,020	2,18**	0,037	0,00	-0,001

7.3 Trend analysis using vegetation indices

Vegetation indices are quantitative measurements indicating the vigour of vegetation. They show a better sensitivity than individual spectral bands for the detection of biomass. Satellite sensor derived NDVI and other related remote sensing data have been successfully used to identify and monitor areas affected by drought at regional and local scales (Bayarjargal et al., 2006). The use of satellite data has become a common process in the quantitative description of vegetation.

The interest of these indices lies in their usefulness in the interpretation of remote sensing images. They constitute notably a method for the detection of land use changes (multitemporal data), the evaluation of vegetative cover density, crop discrimination and crop prediction. An earlier comprehensive review of vegetative indices was performed by Bannari et al. (1995) with Pettorelli et al. (2005) reviewing their usage in assessing environmental conditions.

The NDVI can be used for accurate descriptions of the continental land cover, vegetation classification and vegetation phenology and for the effective monitoring of rainfall and drought, the estimation of net primary production of vegetation, crop growth conditions and crop yields, for detecting weather impacts and other events important for agriculture, ecology and economics. A time series of NDVI images shows the temporal behaviour of the vegetation performance, the vegetation dynamics which can be quantified by applying a rendering analysis to the time series of the NDVI images. The coupling of the NDVI time series to climate parameters, establishes a relationship between the vegetation dynamics and climate simultaneously in spatial and temporal scales (Roerink et al., 2003).

Roerink et al. (2003) used the NOAA AVHRR NDVI for two regions in Europe (1995-1997) and the Sahel (1992-1993) to quantify the spatial and temporal relationships between climate variability and vegetation dynamics. They used the HANTS algorithm (Verhoef, 1996) to derive time series of NDVI images with the number of frequencies set at 3; the mean NDVI (frequency=0), the yearly amplitude (frequency=1), and the amplitude over 6 months (frequency=2), a climate indicator (CI) has been formulated from meteorological data (precipitation over net radiation). A strong correlation was found between the NDVI Fourier components (FC) of the vegetation dynamics and the CI. Zhou et al. (2003) used a statistical analysis to estimate the relation between the

NDVI and climate by land cover type in order to quantify the effects of the climate and other variables on the inter-annual variations in satellite measures of vegetation at the regional scale. The results indicated that temperature changes between the early 1980s and the late 1990s are linked with much of the observed increase in satellite measures of northern forest greenness. They argued that a statistical meaningful relation does not imply causation. Indeed, physical theory indicates that two directions of causality are possible.

There are ecological and physical mechanisms by which climate can affect plant growth and there are physical and ecological mechanisms by which plant growth can affect climate. Davenport and Nicholson (1990) investigated the relation between NDVI and rainfall in East Africa. It was observed that the best relationship is obtained with multi-month rainfall totals and with NDVI lagging rainfall, the highest correlation being with rainfall plus two previous months (Davenport and Nicholson 1990). Kinyanjui (2010) observed rainfall to be very variable in Kenya and to have a major impact on the vegetation status, leading to large changes in NDVI from year to year, and making it quite difficult to identify trends. The study recommended obtaining rainfall data and examining the relationship between NDVI and rainfall.

7.3.1 NDVI and Rainfall

Rainfall data from four stations (Narok, Nyangores, Kiptunga and Bomet) having a good data range for the period (1999-2010) and the NDVI data obtained from SPOT-VEGETATION sensor (§4.2.3.2.1) at the location of the rainfall station, were used to relate NDVI and rainfall. The dekadal composites for the rainfall stations were generated by accumulating the daily rainfall amount in the dekad. The relation between the dekadal rainfall and the NDVI composites is shown in Fig 7-8. There is a weak relation between the dekadal rainfall and the NDVI, in which the scatter plot shows a log-linear response. To improve the relationship between NDVI and rainfall, various studies have processed the daily rainfall into a series of monthly combination. The monthly rainfall has been used in the concurrent month (lag 0, one month earlier (lag 1), two months earlier (lag 2), and three months earlier (lag 3). Combination of (Lag0+lag1, lag0+lag1+lag2, lag0+lag1+lag2+lag3, lag1+lag2+lag3) leading to multi-monthly series have also been used.

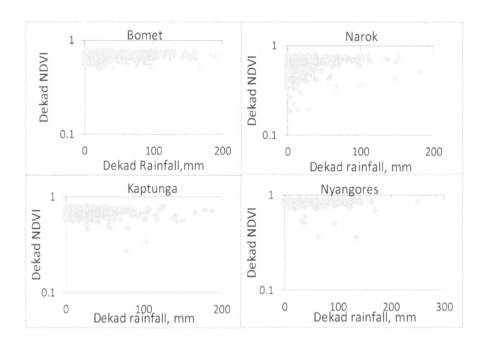

Figure 7-8: Dekadal rainfall and NDVI for the growing period 1999-2008

Davenport and Nicholson (1990) found a high correlation between rainfall and ten different vegetation types in East Africa. The best correlation was in two to three multimonths preceding rainfall in the semi-arid regions of eastern Africa. Hashemi (2011) studied NDVI and rainfall correlation in Azerbaijan. In addition to the monthly and multimonth lagging, he also used the logarithm of multi-monthly rainfall for previous months. The correlation between rainfall and NDVI of different landcover types increased marginal from 0.68 using the preceding two month multimonth data to 0.71 with the logarithm series of the same two-month multi-month. Richards and Poccard (1998) in southern Africa indicated that the accumulation of rainfall (multimonth rainfall) for upto three months gives a high correlation (R^2=0.82). In the Upper Mara basin, both the single one to three months (lag0 –lag3) method, and the accumulation of rainfall (multi-months) method were used to assess the correlation between NDVI and rainfall. All months (lag 0 - lag 3) gives poor correlation between the rainfall and the NDVI (Table 7-5). The best correlation (r=0.4) between the two variables was obtained for three months preceeding the current month in grass landcover type and the Bomet station. The corrrelation between the

NDVI and the preceeding rainfall for the different lag combinations is given in Table 7-5. This apparent poor correlation between rainfall and NDVI in the study area indicates the existence of a more complex relationship that may not be explained by simple linear regressions. The average rainfall in the study area is 1200mm, with some stations averaging as high as 1800mm. This high average annual rainfall in the study area is outside the margins set out by previous studies.

According to Davenport and Nicholson (1990), strong similarity between temporal and spatial patterns of NDVI exists when the annual rainfall is below 1000mm and the monthly rainfall does not exceed 200mm. Above this threshold value, saturation occurs. In a study in southern Africa (south of 15°S) by Richards and Poccard (2010), weak correlations in areas experiencing annual rainfall above 900mm were also reported. According to Davenport and Nicholson, (1990), the saturation response has two interpretations: 1) the constancy of NDVI above this saturation threshold means that at high canopy densities, additional growth has a diminishing influence of photosynthetic activity (assessed directly by NDVI). 2) the saturation response indicates that growth limitations are imposed by other factors rather than rainfall. The NDVI information in this study was generated for the vegetation type at the location of the stations. Due to the considerable impact of human influence on the vegetation and the relatively small sizes of the land holding, the NDVI signal may not be a true representation of either pure land cover type or natural vegetation cover.

7.3.2. Seasonal variation in NDVI (1999-2010)

The study area is generally characterized by bimodal rainfall, with MAM and OND being the wet periods, and JJAS as the long dry period. The MAM rain period shows a fairly stable vegetation response for all the vegetation types. The same stability is also witnessed in the JJA seasons. The OND season has a strong variability for all the vegetation types. A main dip of the NDVI in OND was experienced in 2002, with no noticeable variability in the MAM and JJA periods (Fig 7-9). The NDVI values were higher in MAM than the other periods, surprisingly, and more so for the annual vegetation types, in some years the supposedly dry season JJA has higher NDVI values than even the wet OND season. Trends analysis of the seasonal NDVI shows that all the vegetations types had an upward trend in the accumulated seasonal NDVI (Fig 7-10). Forest and crop cover types had a significant ($\alpha = 0.05$) increase in seasonal NDVI in the MAM and OND seasons respectively.

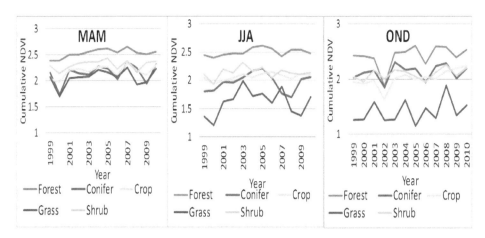

Figure 7-9: Seasonal variation in NDVI for the different vegetation types

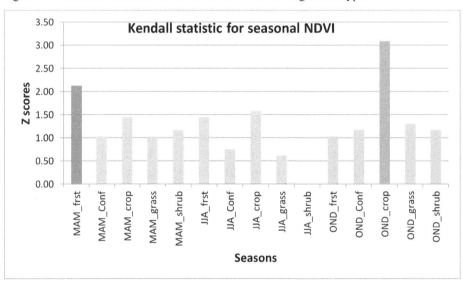

Figure 7-10: Trend analysis for the accumulated seasonal NDVI (1999-2010).

Table 7-5: NDVI and rainfall correlation at different lag periods.

Stations

Landcover type	Narok				Kiptunga				Bomet			
	L0	L1	L2	L3	L0	L1	L2	L3	L0	L1	L2	L3
NDVI_forest	0.1	0.0	-0.1	0.1	0.2	0.1	-0.1	-0.1	0.1	0.0	-0.1	0.0
NDVI_conifer	0.0	0.0	0.0	0.1	0.1	-0.1	-0.2	-0.2	0.0	0.0	0.0	0.2
NDVI_crop	-0.3	-0.3	-0.2	0.2	0.3	0.3	0.2	0.1	-0.2	-0.2	-0.1	0.3
NDVI_grass	-0.4	-0.2	0.0	0.4	0.3	0.2	0.1	0.1	-0.2	-0.1	0.1	0.4
NDVI_shrub	-0.3	-0.2	-0.1	0.2	0.3	0.3	0.1	0.0	-0.2	-0.1	0.0	0.3

Landcover type	Narok				Kiptunga				Bomet			
	L0+L1	L0+L1+L2	L1+L2+L3	L0+L1+L2+L3	L0+L1	L0+L1+L2	L1+L2+L3	L0+L1+L2+L3	L0+L1	L0+L1+L2	L1+L2+L3	L0+L1+L2+L3
NDVI_forest	0.0	0.0	0.0	0.0	0.2	0.2	0.0	0.1	0.0	0.0	0.0	0.0
NDVI_conifer	0.0	0.0	0.1	0.0	0.0	-0.1	-0.2	-0.1	0.0	0.0	0.1	0.1
NDVI_crop	-0.3	-0.3	0.2	-0.2	0.3	0.3	0.3	0.3	-0.2	-0.2	0.0	0.0
NDVI_grass	-0.3	-0.2	0.2	0.0	0.3	0.2	0.2	0.3	-0.1	0.0	0.3	0.2
NDVI_shrub	-0.3	-0.3	0.0	-0.1	0.3	0.3	0.2	0.2	-0.1	-0.1	0.2	0.1

7.3.3 Weather and ecological components of NDVI

To assess the environmental variability of the NDVI in plant productivity, the ecological component was separated from the weather effect by extracting the minimum and maximum NDVI values for each pixel and dekad. The capacity for plant communities to produce relatively high biomass amounts depends more on the total availability of heat and moisture over the same year, than on the conditions during one month of that year, or the anomalous conditions of the previous year. Several annual based indicators have been used to test this hypothesis and the results presented in Table 7-6.

The NDVI annual integral (NDVI-I) is the annual integral of the NDVI calculated by summing up the products of the historical longterm mean NDVI for each period and the proportions of the year represented by that date Guerschman et al., (2003). The NDVI-I is a good estimator of the fraction of the photosynthetic active radiation absorbed by the canopy

$$NDVI - I = \sum_0^n NDVIi \; X \; Ti \; \text{------------------------------------}7\text{-}1$$

Where; n is the total number of composites per year, $NDVI_i$ is the ith composite and T_i is the proportion of the year covered by the ith composite.

This is differentiated from the integrated NDVI (iNDVI) which is the sum of positive NDVI values over a given period mostly annual, which is assumed to be a good indicator for the general land performance (Klien and Roehring, 2008).

$$iNDVI = \sum_0^n NDVIi \; \text{---}7\text{-}2$$

The iNDVI values obtained for the study area are relatively high for all the land use types. This implies a healthy ecosystem which has generally little environmental stresses. Klein and Rohlinger (2003) found comparable, albeit lower iNDVI values for similar vegetation types in the transition zone from the semi-humid to the semi-arid climate of the Laikipia plains in Northen Kenya (Table 7-6). The higher iNDVI values in the Mara are due to the high rainfall experienced in the humid climatic zone. However, the grass land in the Mara study area is located in the semi-arid regions and thus the iNDVI for grassland is lower than that of the transition region in Laikipia. The relative range of the NDVI (RREL) proposed by Guerschman et al. (2003) corresponds to the difference between the maximum and minimum NDVI recorded in the year divided by the NDVI-I. Subtracting the minimum NDVI from the maximum is important since it allows for the separation of the variability of the NDVI related from the contribution of background noise caused by bare soil or geographic resources in the estimation of weather induced impacts (Singh, 2003).

$$RREL = \frac{NDVImax - NDVImin}{NDVI-I} \quad \text{-------------------------------------7-3}$$

The higher RREL for grass shows a big difference between the maximum and minimum NDVI. During the wet season, the grass has full vegetative health and therefore high NDVI. In contrast, during the dry season, the grass dries out, leading to very a low NDVI. The integral NDVI is also low compared with the other land use types.

Table 7-6: Selected indicators for annual NDVI analysis

Landuse type	Indicators							
	iNDVI Mara		iNDVI Laikipia*		NDVI-I		RREL	
	Mean	SD	Mean	SD	Mean	SD	Mean	SD
Forest	29.93	1.11	26.2	4.6	8.31	0.31	1.57	1.91
Conifer	25.09	1.80	-	-	6.97	0.50	2.17	1.71
Crop	24.83	1.56	20.8	11.1	6.90	0.43	1.74	1.15
Grass	20.33	2.84	21.6	8.3	5.65	0.79	5.74	2.44
Shrub	26.13	1.60	18.7	15.5	7.26	0.44	0.88	0.59

*(Klein and Rohlinger, 2003)

Over the 12 years between 1999 and 2010, the general tendency for all the vegetation types is an upward trend, leading to an increase in the annual NDVI (Fig 7-11a). Although the duration under consideration is not long enough to draw a firm conclusion on the trend analysis, the general tendency is that the vegetation is growing healthier (Fig 7-11b). This may be due to increased carbon input leading to more photosynthesis in the natural vegetation or the introduction of better farming husbandry in the agricultural area, leading to more biomass production in crop vegetation types. The increase in the minimum temperatures to near required base temperatures for most vegetation types, especially in the upstream sections of the basin, indicates the increased potential for plant to grow all year round.

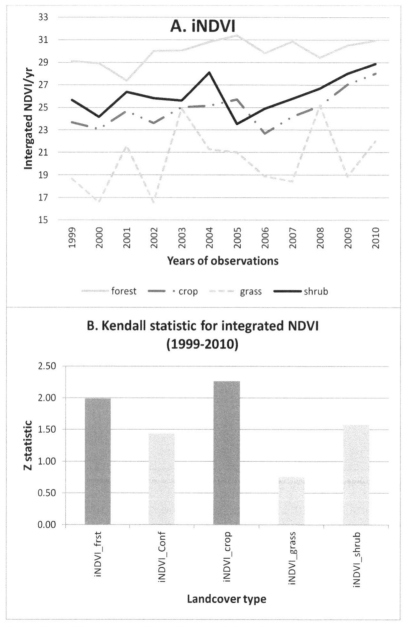

Figure 7-11: Time series (A) and trends (B) in the annual accumulated vegetation health for different land cover classes over a 12 yrs period, crop and forest landcover types significant change (a = 0.05 level of significance)

The success of using NDVI to identify stressed and unhealthy or damaged crops and pastures maybe be replicated in non-homogenous areas where differences between the level of vegetation can be related to differences in environmental resources (Singh et al., 2003). In vegetated regions there is difficulty in detecting weather related NDVI fluctuations, since integrated area of the weather component is smaller than the ecosystem component. There is therefore, need to separate the weather from the ecosystem components of the vegetation. Commonly used indices that may perform the component separation include the Vegetation Condition Index (VCI), the Standard Vegetation Index (SVI) and the Vegetation Productivity Indicator (VPI).

7.3.4 Vegetation condition index (VCI).

The VCI is a normalization of the NDVI for each pixel based on minimum and maximum NDVI values overtime Kogan (1990, 1995).

$$VCIj = \frac{NDVIi - NDVImin}{NDVImax - minNDVImin} * 100 \text{-----------------------------7-4}$$

Where: VCI is the modified NDVI (expressed in %); NDVI is the smoothed dekadal composite Normalized Difference Vegetation Index; minNDVI and maxNDVI are absolute minimum and maximum, respectively, of the smoothed dekadal composite NDVI defined from historical data; i and j define dekad and location, respectively.

In order to analyse the VCI, the greenest month in the year was found by using monthly averages for the 12 years under consideration. For the selected vegetation types, the month of May was found to have the highest NDVI for all except one of the vegetation types (Fig 7-12). Since the VCI is a measure of the vegetation condition, selecting the wettest month (best case scenario) ensures that all other dekads are always below this condition and therefore effectively indicates vegetation stress caused by weather variables, notably precipitation and temperature. The VCI calculations were performed by use of a tool in ERDAS to identify minimum and maximum NDVI in a composite file containing the daily NDVI in a given year (Annex 6a).

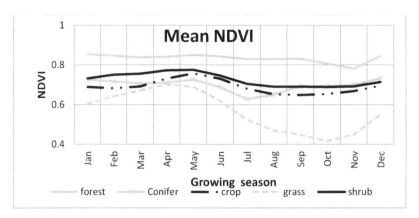

Figure 7-12: Long term monthly NDVI averages for different vegetation types

The results are presented in five classes: 0-20% very poor, 20-40% poor, 40-60% average, 60-80% good and 80-100% very good. The last MVC for the dekad in May (denoted 21/05) for every year was analysed with the VCI formula using the ERDAS tool. The spatial representation of the calculated VCI for the 3rd dekad in May is given in (Fig. 7-13). From the VCI diagrams, 2000 was the driest year in the period 1999 to 2010; this is confirmed in the MAM seasonal NDVI diagrams where 2000 shows a sharp drop in the NDVI. A look in the seasonal rainfall pattern for the overlapping period (1999-2010) shows that the lowest annual rainfall was recorded in 2000 in all the rainfall stations within the study area(Fig.7-14). This implies that the VCI has successfully managed to separate the ecosystem noise from the weather input signal.

7.3.5 Standard Vegetation Index

The standard Vegetation Index (SVI) was proposed by (Peters et al., 2002) and is relies on the understanding that vegetation conditions are closely linked to the weather conditions in the atmosphere closest to the ground. It shows the effects of climate on vegetation over short-time periods. The SVI is based on calculation of a z score (for normally distributed parameter), otherwise a t-score for each NDVI pixel location in the study area. The z (or t) score is a deviation from the mean in units of the standard deviation, calculated from the NDVI values for each pixel location for each dekad for each year, during the years 1999-2010:

$$Z_{ijk} = \frac{NDVI_{ijk} - meanNDVI_{ij}}{6_{ij}}$$ ---7-5

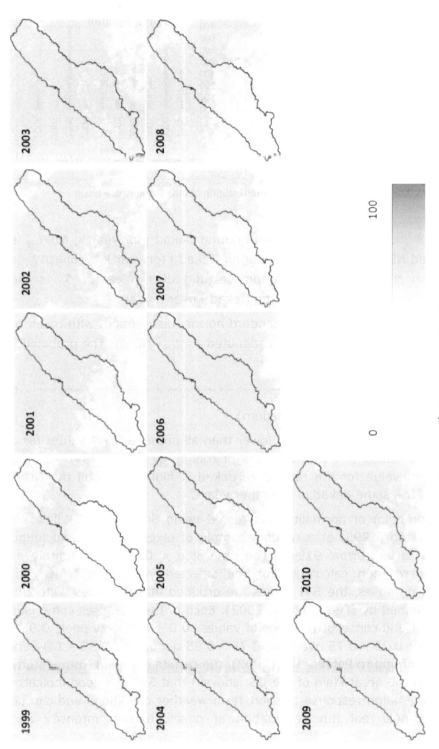

Figure 7-13: VCI (0= very poor and 100= very good) for the 3rd dekad in May (21/05).

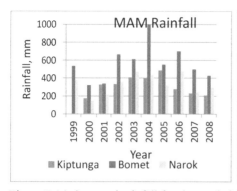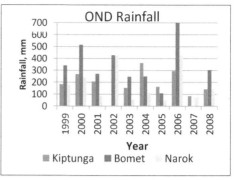

Figure 7-14: Seasonal rainfall for three rainfall stations in the Upper mara basin

Where: Z_{ijk} is the z-value for pixel i during Dekad j for year k, $NDVI_{ijk}$ is the Dekad NDVI value for pixel i during Dekad j for year k, $meanNDVI_{ij}$ is the mean NDVI for pixel i during Dekad j over n years, σ_{ij} is the standard deviation of pixel i during Dekad j over n years.

The Z_{ijk} was assumed to fit a standard normal distribution, with mean of zero and standard deviation of 1, denoted as Z_{ijk}- N(0,l). The probability density function of Z_{ijk} is given by:

$$Z_{ijk} = prob(Z_{ijk} < Z_x, \text{V}) \text{--- 7.6}$$

Where: v = n-1 (degrees of freedom)

If SVI=0; a pixel NDVI value is lower than all possible NDVI values for a pixel for a dekad in all other years of this study and for SVI =1; the pixel NDVI value for the respective dekad is higher than all the NDVI values of the same dekad in the other years.

The assumption of normality was tested using Shapiro and Wilk's W-test, Royston (1995) at a random sample of pixel locations and found the data to be normal 91% of the time at a = 0.01. The monthly p-values were then calculated for the selected vegetation types. For mapping purposes, the SVI values are grouped into 5 classes, with the ranges defined by (Peters et al., 2002). Each of these classes comprises a different and consecutive range of values: 0.0 - 0.05 very poor; 0.05 - 0.25 poor; 0.25 - 0.75 average; 0.75 - 0.95 good; and 0.95 - 1.0 very good. According to Peters et al (2002), the results obtained from a study applied to the great Plain of the US, showed that SVI is a good indicator of the vegetation response to short term weather conditions and can be used for near-real time evaluation of onset, extent, intensity and duration of vegetation stress. The mean and standard deviation functions in modelmaker were scripted to calculate the mean NDVI and the standard deviation for each dekad in the time series (Annex 6b). As per the SVI formula, the z score was calculated by subtracting the mean from the current NDVI and dividing the result by the standard deviation.

The z scores obtained from the formula as given by Peters et al. (2002) were converted to P-values (SVI) assuming normal distribution with a mean of zero and a standard deviation of one, the results were grouped into the five groups: very poor, poor, average, good, very good. The distribution of vegetation conditions into these classes was intended to mimic a normal probability density function. For instance, a pixel that is classified as "very poor" indicates that its NDVI value is lower than the average during the same week of the year relative to that in other years of the study.

A pixel classified as very good indicates that its NDVI value is higher than average, or that vegetation is in very good relative condition. For all the vegetation types, most of the data values lay in the average class. The forest vegetation type has the narrowest normal distribution fit with most of the dekads in the average class and none in the extreme very poor or very good classes. This indicates minimal variability over the entire study period. For all the vegetation classes, the tendency is skewness in distribution towards the good and very good classes. The vegetation health has experienced a relative better than average condition (Fig 7-15).

Several shortcomings were identified in the VCI and SVI methods. VCI assumes that the current range represents a maximum possible variation and that all values of NDVI within the range occur with an unrealistic samilar frequency (Sannier et al., 1998). SVI assumes normal distribution of the NDVI series. However, the true parameters of the NDVI population is unknown given the short duration of the available data. This uncertainty has been accounted for by using the students-t distribution that has a wider "spread" than the normal distribution.

7.3.6 Vegetation productivity indicator (VPI)

Developed by Sannier (1998), the method empirically estimates the statistical distribution of NDVI from the available data without any assumption and is sensitive to the background vegetation type. All pixels in the time series are ranked from the lowest to the highest. The probability of occurrence over the time period is given by:

$$P = \frac{m}{n+1}$$ 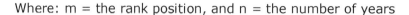 7-6

Where: m = the rank position, and n = the number of years

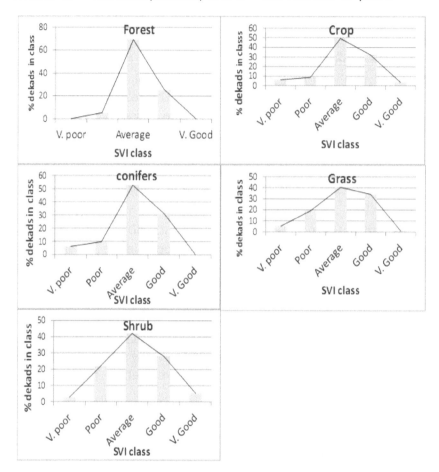

Figure 7-15: SVI classes distributions for vegetation types in the Upper Mara basin with enveloping of the distribution

A low P indicates that it is unlikely to get a lower NDVI value and thus corresponds to poor conditions; a high P indicates better than norm conditions.

The Vegetation Productivity Indicator (VPI) is used to assess the overall vegetation condition and is a categorical type of the difference vegetation index, whereby the actual NDVI is referenced against the NDVI percentiles of the historical year. The VPI is classified in 5 groups: 0-20% is very low, 21-40% as low, 41-60% average, 61-80% as high, and 80-100% as very high (Fig 7-16). The health of the vegetation has been considerably good over the study period.

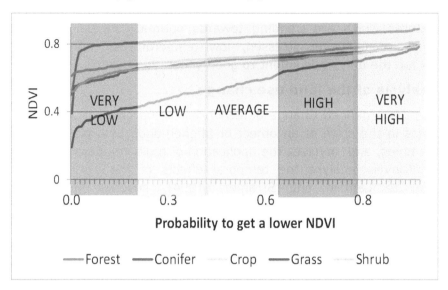

Figure 7-16: VPI classes for the different vegetation types in the Upper Mara.

7.4 Relating trends in vegetation and climatic variables

The weak correlation between the rainfall and the NDVI has indicated that primary productivity in the study area is not very sensitive to the rainfall variability. The area recieves rainfall amounts way beyond the threshold levels. The trend analysis indicates declining tendency in rainfall amounts over most of the stations in the study area (see § 7.2.1). At the average decline rate in annual rainfall of 6mm/yr, it will take more than 30yrs for the rain average basin rainfall to fall to the threshold annual value of 900mm. This is without considering climate change effects, many GCMs predict a wet climate in this eastern African highland area (IPCC, 2007).

The minimum temperature has significantly increased in the last 20 yrs. This increase in minimum temperature has a positive effect on the vegetation. Temperatures govern annual productivity in various ways that do not result from temperature dependence of the photosynthetic process (Lieth, 1973). The optimum temperature for productivity, in the range of 15-25°C, agrees with the optimum temperature range for photosynthesis. Photosynthetic productions follow the van't Hoff equation, with productivity doubling every 10°C between the temperatures of-10 and 20°C. The increase in minimum temperature leads to higher primary production and therefore vegetation health. The sustained increasing trend in the vegetation as indicated by all the vegetation indices is consistent with expectations under the minimum temperatures which are trending towards optimal ranges. Since the vegetation is not water stressed, low temperatures, nutrients and diseases are the only impediment to good vegetation health.

7.5. Analysis of the land use change

Lu et al., 2004 defined Change detection as "the process of identifying differences in the state of an object or phenomenon by observing it at different times, and involves the application of multi-temporal data sets to quantitatively analyse the temporal effects of the phenomenon". According to Lambin and Strahler (1994), several natural and anthropogenic factors influence changes in the land cover. Natural factors include changes and variability in climatic conditions, ecological, and geomorphological processes, while anthropogenic factors include deforestation, land degradation and human related greenhouse gas emissions. The main characteristics of a good change detection study should include and not limited to: areal and rate of change analysis, spatial extent of change analysis, trajectories in the changed cover types and an analysis of the change detection accuracy (Lu et al. (2004).

For successful implementation of the change detection, remote sensed data should be consistent. This may be achieved by using same sensor, same radiomentric resolution, same spatial extent at anniversary or close to anniversary acquisition dates. This consistency minimises the effects of errors due to sun angle and phenological differences caused by seasonal changes (Jensen, 1996). Though the images used in this study were not from the same sensor, there are little differences between the MSS, the TM and the ETM+ sensors. All the images utilised in this study were acquired dated late January to early February (quasi-anniversary).

The 1976 MSS image with 80 m pixel resolution were resampled to 30 m to match the spatial resolution of the optical bands of the TM and ETM+ data. Three methods were used to analyze the evolution of land changes in the study area. Firstly, an indirect quantitative postclassification method which involved the calculation of the area covered under different land cover types and comparing with the corresponding landcover type from one time period to another. This method is independent of any change detection algorithm and accuracy of the change detection is only subject to the quality of the classified thematic map.

Secondly, the digital change detection of change/non-change information based on the image differencing method which is a binary method where registered images acquired at different times are subtracted to produce a residual image which represents the change between the two dates at a preset threshold value (Jensen, 1986). The critical aspects are both the selection of suitable image bands and the selection of suitable thresholds to identify the changed areas. Thirdly, a RGB-NDVI change-detection method involving the vegetation index differencing, with combinations of the primary (RGB) or complimentary (yellow, magenta, cyan) colors (Sader et al., 2001).

7.5.1 Postclassification method of change detection

In this method, multi-temporal images are separately classified into thematic maps, then a comparison of the classified images, pixel by pixel is performed. The main advantages of the method are: it minimizes impacts of atmospheric, sensor and environmental differences between multi-temporal images, and it provides a complete matrix of change information. The disadvantages are: it requires a great amount of time and expertise to create classification products, and that inaccuracies in two date classifications can be multiplicative (the final accuracy depends on the quality of the classified image of each date) (Singh, 1989; Lu et al., 2004). Also, the method does not allow for normalizing differences between multi-temporal data (Muchoney and Haack, 1994).

The area in the thematic maps under the various land cover classes was determined by use of geo-spatial tools. The earlier of the two years under consideration was used as a reference and the latter year is the year upto which the change occurs. The change in one landcover over a period was calculated by substracting the value in the reference year from the final year. Fig. 7-17 presents the results of the change detection over the three decades. Overall, in the three decades (1976-2006), only the area under crop cover increased in size (109%), while the other cover types all declined at varying degrees. Forest cover decreased by 27720 Ha, shrub cover by 24508 Ha, and grassland by 5000 Ha, representing a change rate of 34%, 31% and 4%, respectively.

The different cover types experienced rapid changes in their cover at different decades. The changed area and the spatial extend of the changes are presented in Fig. 7-18. The 70s (1976-86) saw the rapid decline in grass cover as people settled on the flat plains. At the same time there was the expansion of privately and government owned plantation forests in the upper parts of the basin.

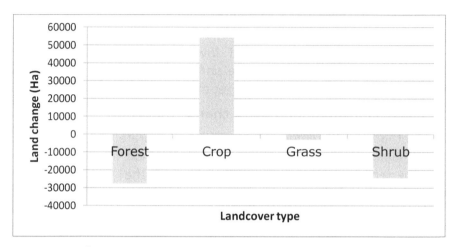

Figure 7-17: Changes in the landcover over three (3) decades (1976 to 2006).

The destruction of shrubs and trees in the lower areas for firewood, charcoal and construction material led to the significant decline of the shrubland cover in the 80s (1986-1995). The excision of forested land for farming and the introduction of the tea growing and forest agriculture in the 90s (1995-2006) saw a decline in the forest cover. According to Obare and Wangwe (1998), "In the 1990s the Forest Department (FD) introduced the Non-Resident Cultivation (NRC) for the establishment of plantation forests". In the 1940s, the *Shamba* system was introduced to facilitate plantation establishment. It was prompted by the acute land shortage faced by communities after colonization, a need to reduce plantation establishment costs by the Forest Department, and to provide food security to those who practiced it.

Under the *Shamba* system, the cultivators were incorporated into the FD through employment and were permitted to clear and cultivate cut over indigenous bush cover from a specified land area; usually between 0.4-0.8 ha per year. This is done with the agreement that tree seedlings are planted on this land, and subsequently tended through weeding, pruning and safeguarding against game damage. In return, the FD provided the resident cultivator with employment, social amenities and land for the cultivation of annual crops such as maize, potatoes, beans, peas and other vegetables. Cultivation proceeded until a time when tree seedlings were large enough to shade, and thus inhibit the growth of annual plant crops; usually a period of 3-5 years (Obare and Wangwe, 1998).

The NRC was therefore a modification of the *Shamba* system that attempts to reduce the risk of cultivators claiming squatter rights on forest land. Besides the officially degazetted forest, encroachment is a major problem in the Mau forest area (Table 7-7) The introduction of extensive production of grass crops (wheat and barley) has lead to the increased grass coverage in the decade investigated (1995-2006)

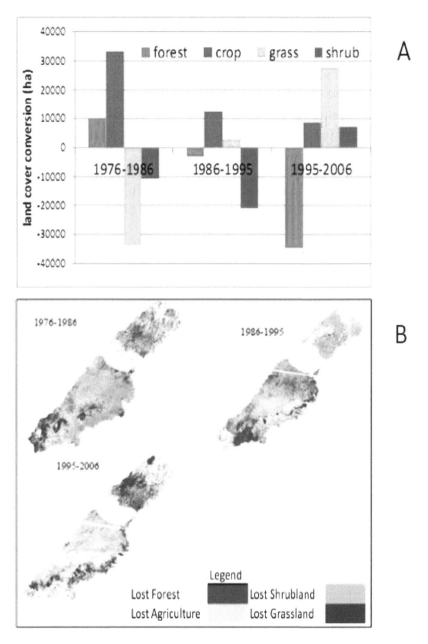

Figure 7-18: Decadal change analysis based on classified images. The area lost or gained by the different land types (A) in each of the three decades, and the spatial extent of the loss/gain in the Upper Mara basin(B).

Table 7-7: Causes of encroachment in East African forest landscapes (Banana et al., 2010).

	Cause	Description
1.	Fertile forest soils	The relatively rich and virgin forest soils attract encroachers because they employ poor farming methods and seriously degrade and exhaust soils outside forest reserves. However, this soil also gets leached much faster when exposed to high temperatures and heavy tropical rainfall, and is quickly exhausted. This forces encroachers to open up new land annually and hence to clear more forests.
2.	Breakdown in law enforcement	For a long time, the Forest Department (FD) staff in the region has not been able to enforce the law. Patrols have been intermittent and at large intervals, which enables encroachers to settle unchallenged in forest lands.
3.	Unclear forest boundaries	Many local communities that are adjacent to forests have crossed the boundaries, knowingly or unknowingly, since the boundaries are unclear. Many boundary marker shave been destroyed and the positions of others altered by encroachers to obfuscate where the boundary is.
4.	Corrupt government officials	Some corrupt forest officials, in connivance with other relevant officials, encourage encroachment
5.	Breakdown in monitoring permits	Forest officials allow some activities, like grazing and construction of temporary structures in central forest reserves (CFRs), on renewable permit terms. Over time, many of these people fail to renew their permits but continue with their activities.
6.	Apparent shortage of land outside CFRs.	In some cases, local population pressure has pushed people into adjacent CFRs.
7.	Encroachment in CFRs	Often, encroachment in CFRs has the backing of politicians, who usually trade CFR land for votes. Very often, local leaders are themselves encroachers and, when faced with eviction, tend to exaggerate the number of encroachers to enhance their stakes and win the sympathy of the public/government.
8.	Lack of awareness of government policies and laws	Quite often, encroachers are not aware of the policies and laws on forestry.

7.5.2 Change detection by image differencing

Image differencing is an algebraic method based on selecting thresholds to determine the changed areas, but do not provide complete matrices of change information. One or more wavebands from two, co-registered images are subtracted to produce a residual image indicating the relative change in reflectance between the two dates. The approach detects all changes greater than the pre-set thresholds and provides detailed change information. The challenge is in the criteria to be used to select suitable thresholds to identify the changed areas, which is subjectively left to the devices of the analyst (Lu et al., 2004). Despite the analysis of three or more dates of imagery allowing for trends to be examined at more than one interval of time, the common practice is for digital change detection is to be performed stepwise using only two dates of satellite imagery at each step.

Fig 7-19 shows the differences in the two sets of satellite images and the highlighted differences using a ±10% threshold. During the 1986-1995 decade, there are lots of changes in the vegetation, both in terms of vegetation gain and vegetation loss. The vegetation gain is the transition from a vegetation type with a lower greenness to one with a higher one. The change from grass to crops like tea will mean there is higher NDVI capture. The conversion of fallow grassland to agricultural land with periods of bare land during field preparation (as is the case in February) will leave land surface bare and exposed. This will be reflected in the image difference as a vegetation loss. Similarly the cutting down of perennial landcover types like shrub through activities like charcoal burning leaves the ground bare with minimal NDVI and will thus appear as a vegetation loss in the image difference.

Although care has been taken in the acquisition of the satellite images used to develop NDVI by using near anniversary dates images, other unmitigated natural occurrences do affect the outcome. The seasonal variability in climate and especially precipitation amounts and timing may have significant impact on the derived NDVI.The visual RGB-NDVI method involves creation of color composite images and utilization of the additive color theory, where each NDVI from three dates is combined with the red, green, and blue color write functions of the computer monitor (Sader and Winne, 1992, Sader et al., 2001, Sader et al., 2003). To automate the change detection and turn the three-layer NDVI stack into a thematic map, an unsupervised classification (ISODATA clustering) is performed on each NDVI stack (i.e 1976-1986-1995 and 1986-1995-2006).

The RGB-NDVI change-detection method was found by Hayes and Sader (2001) to be the most accurate and efficient for a tropical forest study site when compared to the image differencing, and the principal component analysis methods. When determining change, RGB-NDVI incorporated three dates at one time as opposed to two-date stepwise sequences using image differencing methods. This is particularly advantageous when several image dates are analyzed in a sequence to detect change. The RGB - NDVI unsupervised classification avoided analyst subjectivity in selecting appropriate histogram thresholds for several stepwise sequences and was more straight-forward and time efficient. Interpretation of change over time was intuitive and logical using additive color theory. Finally, the interpretation of the clusters and their multivariate statistics facilitated identification of forest clearing, no change, and regrowth classes in a time-series.

The RGB-NDVI method was used to make two groups of three dates at a time (1976 to 1995; 1986 to 2006). By simultaneously projecting each of the three NDVI dates through the red, green, and blue (RGB) computer display write functions, major changes in NDVI between dates will appear in combinations of the primary (RGB) or complimentary (yellow, magenta, cyan) colors. In Fig 7-20, the three layers comprising the three dates that need to be compared are stack together. Sequentially the oldest to the most recent dates are used as band red, green and blue respectively.

The magnitude and direction of vegetation cover can visually be interpreted by knowing which date of NDVI is coupled with each display color. Changes in the study area over the three dates automated classification was performed on three or more dates of NDVI by the ISODATA unsupervised cluster analysis technique.

Change and no change categories are labeled and dated by analysis of the cluster statistical data and guided by visual interpretation of RGB-NDVI color composites. Table 7-8 shows the interpretation of the additive colours in the three dates RGB-NDVI. Cluster busting using a 7 x 7 majority window was performed to segment clusters between change and no change and then re-clustered to reduce confusion and to better distinguish between vegetation gains and vegetation loss areas (Гig. 7 21).

The extent of the gain/loss in each decade is given in Table 7-9. In the first decade (1986-1995), the amount of vegetation cover lost was slightly lower than that gained. However, in the 1995-2006 decade large areas under vegetation cover were lost. This was especially experienced in the areas around the forest areas where large biomass vegetation like trees and forest cover were replaced with annual crops which exposed the bare ground in the dry months of January/february. In the lower areas of the catchment, the loss of savanna grassland by both overgrazing and subsistence cultivation led to a high loss of vegetation cover.

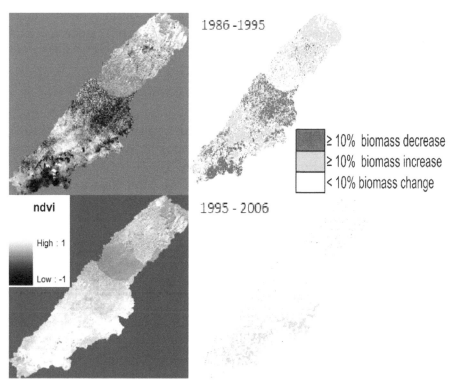

Figure 7-19: Binary image differences for a 10% threshold (left) and the highlights of the dfferences (right) in the NDVI images for 1986-1995 (top) and 1995-2006 (bottom).

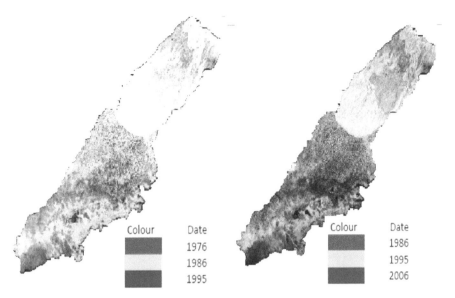

Figure 7-20: RGB-NDVI images for two time periods, 1976, 1986, 1995 (left) and 1986, 1995, 2006 (right).

Table 7-8: Interpretation of RGB-NDVI image (modified after Sader and Winnie, 1992).

Display color Gun Image color	NDVI date RED NDVI 1986	GREEN NDVI 1995	BLUE NDVI 2006	Interpretation of vegetation change
Red	High	Low	Low	Vegetation loss, 1986-1995
Green	Low	High	Low	Vegetation gain, 1986-1995
Blue	Low	High	High	Vegetation gain, 1995-2006
Yellow	High	High	Low	Vegetation loss, 1995-2006
Grey	Low	low	Low	No change, low biomass
Grey	High	High	High	No change, high biomas

Table 7-9: Vegetation trajectories based on the RGB-NDVI method

	Trajectory	Area (Ha)
1	Vegetation loss 1986-1995	51059
2	Vegetation gain 1986-1995	87751
3	Vegetation gain 1995-2006	1944
4	Vegetation loss 1995-2006	109848
5	No change	45070

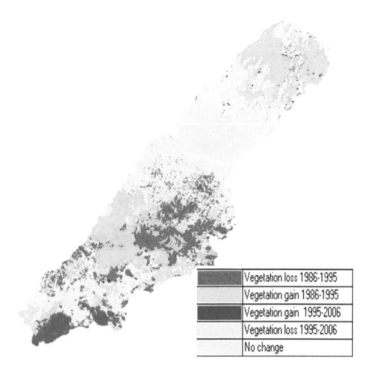

	Vegetation loss 1986-1995
	Vegetation gain 1986-1995
	Vegetation gain 1995-2006
	Vegetation loss 1995-2006
	No change

Figure 7-21: Vegetation change trajectories for two decades 1986-2006 using the three date RGB-NDVI.

The trajectories of the landcover change for the two independent methods point to consistent conclusion in the way land cover changes have taken place in the study area. The decades have been characterized by different changes occurring at different specific regions within the study area. These changes have been driven by both policy changes in the administration of the natural resources as well as demographic and economic pressures. The uncertainties in the methods are from different sources. In the post classification methods, errors are subjective due to modeler's shortcomings. In the image differencing, raw satellite images are used, objective errors due to sensors inadequacies are experienced.

7.6 Conclusion

Generally the rainfall in the Upper Mara basin has a decreasing trend. While five of the six stations have a negative trend only one station has a positive trend. The magnitude of the decrease in rainfall is highest closer to the forest area. This indicates a relation between the forest degradation and deforestation and loss of precipitation. The forested areas have suffered a 34% loss in coverage between 1976 and 2006. Temperature rise has been experinced in all the stations used in the study. The highest rise in temperature was noted in the Narok station which has a grassland/agriculture cover type. The magnitude of change in temperature of $0.04^{o}C$ was consistent with the projected increases for Africa for the last century. The increase of agricultural land (109%) and marginal decrease (4%) in grassland areas, which have the highest land surface temperatures might have contributed to the higher increase in the temperature for this station. The other stations have also been affected by changes land cover especially the introduction of tea farming in the Kericho area and the opulation explosion in the Kisii county.

8. Hydrological modeling

8.1 Introduction to hydrological modeling

A hydrological model is typically a process-based, continuous, dynamic simulator that is based on mathematical descriptions of physical, biogeochemical and hydro-chemical processes by combining elements of both physical and conceptual, semi-empirical nature and includes a reasonable spatial disaggregation scheme (Krysanova et al., 2008). The two main objectives of model applications are: (i) understanding processes and (ii) scenario analysis. Understanding processes is the basis for model software development. In order to build modeling software, the modeler must have a clear picture on how processes in the real world function and how these processes can be mimicked in the model code. The main challenge is not in trying to build-in all the understood processes, which is in fact impossible, but lies in the capabilities to simplify things and concentrate on the most relevant processes of the model under construction. Selecting an appropriate simulator from a pool of so many "useable" simulators is always a big challenge. Two commonly used criteria are the spatial scale to be incorporated in the study and how much physical detail needs to be included, i.e the space-quantity continuum in hydrological modeling.

Different modeling approaches and methods have been used with different assumptions to derive the potential impact of a changed climate on the river basin discharge. There is no accepted method or approach for a proper assessment of global changes, although using different models, assumptions and methods can lead to different conclusions regarding the impact of climate change on water resources (Yates, 1996). Large-scale and complex environmental systems such as the global hydrological cycle or the water quality in a river basin cannot be investigated directly through experimentation, but instead must be generalized into their component processes (Praskievicz et al., 2009). According to Melone et al. (2005), the primary features for distinguishing hydrological modeling approaches include: i) the nature of the algorithms, empirical, conceptual or physically-based, ii) parameter specification, stochastic or deterministic approach iii) the spatial representation, lumped or distributed.

A basin scale model simulates hydrologic processes by fully incorporating the watershed area, a physically based model utilizes physically based equations to hydrological proceeses. A semi-distributed model partially allows the hydrologic processes, input, boundary conditions and watershed characteristics to vary in space by dividing the basin into a number of smaller sub-basins, which in turn are treated as a single unit (Daniel et al., 2011).

According to Borah and Bera (2003), the more commonly used watershed-scale models are: Agricultural Non-point Source (AGNPS, Young et al., 1987, Annualized Agricultural Non-point Source (AnnAGNPS, Cronshey and Theurer, 1998)), Area Non-point Source Watershed Environment Response Simulation (ANSWERS/ANSWERS-2000, Bouraoui and Dillaha, 1996), Gridded Surface Hydrologic Analysis (GSSHA, Downer and Ogden, 2006), Hydrologic Engineering Center's Hydrologic Modeling System (HEC-HMS, USACE, 1995), Hydrological Simulation Program – FORTRAN (HSPF, Bicknell et al., (2001), KINematic Runoff and ERO-Sion (KINEROS2, Goodrich et al., 2002)), Systéme Hydrologique Européen (MIKE SHE, Abbott et al., 1986; DHI, 1993)), Precipitation-Runoff Modeling System (PRMS, Leavesley, 1983)), Soil and Water Assessment Tool (SWAT, Arnold, (1998)) and Water Erosion Prediction Project (WEPP, Laflen et al., 1991). AnnAGNPS, HSPF, MIKE SHE, and SWAT are long-term continuous simulation models that consider the three major components (hydrology, sediment and chemical) that are applicable to watershed-scale catchments (Borah and Bera, 2003).

The AnnAGNPS is suited for agriculture watersheds and is widely used for evaluating a wide variety of conservation practices and other BMPs. It uses a daily or sub-daily time step and it divides the basin into homogeneous land areas, reaches and impoundments. Cropping systems, fertilizer applications, water and dissolved nutrients from point sources, sediment with attached chemicals from gullies, soluble nutrient contributions from feedlots and the effect of terraced fields can be modeled. Some shortcomings of the simulator include: Single day routing of all runoff and associated sediment, nutrients and pesticides loads, lack of mass balance calculations tracking inflow and outflow of water, requirement for an additional data input tool to ease the burden of developing input data sets for the model (Bosch et al., 1998).

The HSPF is a lumped parameter hydrologic model that can simulate the primary natural hydrological processes. The unsaturated zone is approximated using a single storage reservoir. The explicit representation of vegetation in the simulator is limited. The model runs at any time step, from 1 minute to 1 day. The model empirically simulates evapotranspiration from the interception storage, the upper and lower zone storages, the active groundwater storage and directly from the basefow (Bicknell et al., 2001). The HSPF is generally used to assess the effects of land use change, reservoir operations, point or non-point source treatment alternatives, and flow diversions and is also suitable for mixed agricultural and urban watersheds (Daniel et al., 2011). According to Beckers et al. (2009), the HSPF simulator is parameter intensive and offers little advantage with respect to parameterization and calibration over physically based modeling software. Its complexity is compounded by its user-unfriendliness, which makes it difficult to use without direct guidance from an experienced HSPF model user. Furthermore, many parameters that control the hydrologic processes are empirical and can only be determined through calibration.

MIKE SHE is a physically based simulator using multi-dimensional flow-governing equations with numerical solution schemes, which make the models computationally intensive and subject to numerical instabilities. MIKE SHE is mainly suited for small watersheds, for detailed studies of the hydrology and nonpoint-source pollution under single rainfall events or for long-term simulations (Borah and Bera, 2003).

The Soil and Water Assessment Tool (SWAT) model is a physically based, deterministic, continuous, basin scale, semi-distributed hydrological simulator (Arnold, 1993; Gassman et al., 2007). As a physically based simulator, SWAT requires specific information about weather, soil, topography, vegetation and land management practices rather than incorporating regression equations to describe the relationships between input and output variables

8.2 The Soil and Water Assessment Tool (SWAT)

The Soil and Water Assessment Tool (SWAT) is a dynamic, long-term, distributed parameter model (Arnold et al., 1998) with applications in watersheds having agriculture as the primary land use (Manguerra and Engel, 1998). The simulator subdivides a watershed into subbasins connected to a stream network and further delineates the subbasin into hydrologic response units (HRUs) consisting of unique combinations of land cover, soil and topographical slope. The model assumes that there are no interactions among the HRUs and these HRUs are virtually located within each subbasin (Yang et al., 2007). The model calculations are performed on a HRU basis and flow and water quality variables are routed from HRU to subbasin and subsequently to the watershed outlet. The hydrologic routines within SWAT account for vadose zone processes (including inflitration, evaporation, plant uptake, lateral flows and percolation) and for groundwater flow. The SWAT model simulates the hydrology as a two-component system, comprised of land hydrology and channel hydrology. The land portion of the hydrologic cycle is based on a water mass balance. The soil water balance is presented in Eqn 8-1 (all terms mmH$_2$O) as:

$$W_t = W_0 + \sum_{i=1}^{t}(P_i - ET_i - SQ_i - LQ_i - RQ_i - WUS_i - WUD_i - DL_i)\text{-----------}8.1$$

Where: W_t is the water content of the land phase at time t (day), W_0 is the initial water content of the land phase, P_i is the amount of precipitation on the day i, ET_i is the amount evapotranspiration on day i, SQ_i is the amount of surface runoff on day i, LQ_i is the amount of lateral flow on day i, RQ_i is the amount of return flow on day i, WUS_i is the amount of water that is pumped out of the shallow aquifer for external use of the day i, WUD_i is the amount of water that is pumped out of the deep aquifer for the external use on day i, and DL_i is the amount of water that is lost from the system through the deep aquifer.

Water enters the SWAT model's watershed system boundary predominantly in the form of precipitation. Precipitation inputs for hydrologic calculations can either be measured data or simulated with the weather generator available in the SWAT model. Precipitation is partitioned into different water pathways depending on system characteristics (Fig 8-1). The water balance of each HRU in the watershed contains four storage volumes: snow, the soil profile (0-2 m), the shallow aquifer (2-20m) and the deep aquifer (>20 m). The soil profile can contain several layers. The soil water processes include infiltration, percolation, evaporation, plant uptake, and lateral flow.

Surface runoff volume is estimated using the modified version of the United States Department of Agriculture - Soil Conservation Service curve number method (CNII) or the Green-Ampt infiltration equation. The kinematic storage model is used to predict lateral flow in each soil layer. To account for multiple layers, the model is applied to each soil layer independently, starting at the upper layer.

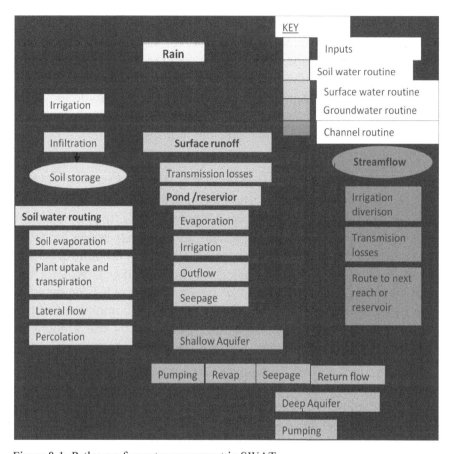

Figure 8-1: Pathways for water movement in SWAT

There are two aquifer systems in SWAT, the unconfined shallow aquifer and the confined deep aquifer (Arnold et al., 1993). The Shallow aquifer contributes return flow to streams within the basin and replenishes moisture in the soil profile in very dry conditions. It may also be directly removed by the plants (Nietsch et al., 2007). The deep aquifer contributes return flow to streams outside the basin boundaries and could also be removed through pumping (Fig 8-11).

Three PET methods, the Penman-Monteith method (Penman, 1956; Monteith, 1965), the Hargreaves method (Hargreaves and Samani, 1985) and the Priestley-Taylor method (Priestley and Taylor, 1972) are incorporated in the SWAT simulators (Arnold et al., 1998). The Penman-Monteith method requires solar radiation, air temperature, relative humidity and wind speed as inputs (Donatelli et al., 2004, Nietsch et al., 2007). The Penman-Monteith method is commonly used (Migliaccio and

Srivastava, 2007), has been described as universally accurate (Wang et al., 2009) and is considered to be the best method when a full complement of weather data is available (Allen et al., 1998). The Hargreaves method requires both maximum and minimum air temperatures as the only input. According to Droogers and Allen (2002), Saghravani et al., (2009), the wide-spread availability of temperatures records in many weather stations has increased the usage of the Hargreaves equation enormously. The Hargreaves method has been shown to overpredict PET in humid climates (Jensen et al., 1990, Amatya, 1995) and to underestimate it in very dry regions (Droogers and Allen, 2002). In the Priestley-Taylor method (Priestley and Taylor 1972) the energy component is replaced by a coefficient (Nietsch et al., 2002) and is reported to perform better in SWAT than the Penman Monteith (Amatya et al., 1995), and the Hargreaves (Lu et al., 2005) for wet and humid surfaces.

Channel flood routing is estimated using the Muskingum method or the Variable Storage Method. The Muskingum routing method accounts for flooding by modeling storage volume in a channel length as a combination of wedge and prism storages that can be expressed as bank storage. Outflow from a channel is also adjusted from transmission losses, evaporation, diversions, and return flow (Zhang et al., 2008). Erosion is estimated using the Modified Universal Soil Loss Equation (MUSLE). After the sediment yield is evaluated using the MUSLE equation, SWAT further corrects this value by considering snow cover effects and the sediment lag in the surface runoff. The sediment routing model that simulates the sediment transport in the channel network consists of two components operating simultaneously: deposition and degradation (Arnold et al., 1995a).

The nutrients simulated in the soil profile of SWAT are nitrogen (N) and phosphorous (P). The soil nitrogen is partitioned into five N pools with two being inorganic (ammonium-N (NH_4^+-N) and nitrate-N (NO_3^--N)) and three being organic (active, stable, and fresh). SWAT simulates the movement between the N pools, such as mineralization, decomposition/immobilization, nitrification, denitrification and ammonia volatilization. Other soil processes such as N fixation by legumes and NO_3^--N movement in the water are also included in the model. Soil phosphorous is divided into six P pools. Three of the pools are characterized as mineral P and three are characterized as organic P. Transformations of soil P between these six pools are regulated by algorithms that represent mineralization, decomposition, and immobilization. Other soil P processes included in SWAT are inorganic P sorption and leaching. The QUAL2E model algorithms (Brown and Barnwell 1987) are used to describe N and P transformations in the channel reaches. Once N enters a channel reach, SWAT partitions N into four pools: organic N, NH_4^+-N, NO_2^--N and NO_3^--N. SWAT then simulates changes in N that results in the movement of N between pools. Two pools of P are simulated for channel processes: organic P and inorganic/soluble P.

The SWAT model simulates on a daily time step, the model has options for the output that allow the user to define the output time step (daily, monthly, or annual). Output variables include flow volume, nutrient yields, sediment yield, and plant biomass yields. These variables are provided on the subbasin or HRU spatial level depending on the output time step selected. The output files generated by the SWAT model are created in text and database file formats.

8.3. Limitations of SWAT modeling in the tropics

The SWAT model has gained widespread application over its more than 30 years due to: i). comprehensive considerations of processes (hydrologic, biological and environmental), ii) inclusion of scenarios, iii) availability of parameter databases iv). Robustness and excellent support from user groups and developers, and V). its open source with good documentation.

Several comprehensive reviews of SWAT have been published, touching on developments, applications and future research opportunities (Kosky and Engel 1997, Arnold and Fohrer 2005, Gassman et al., 2007, Douglas-Mankin et al., 2010, Arnold et al., 2010). Luo et al., (2008) noted that "while SWAT is widely applied to a broad range of conditions, few studies have reported on the variability and transferability of the model parameters and on evaluation of its crop growth, soil water and groundwater modules using extensive field experimental data at the process scale". The local water balance inside the basins is of prime interest in many model applications.

8.3.1 The crop growth module

The plant growth module in SWAT assumes a uniform, single plant species community (Krysanova and Arnold, 2008; Kiniry et al., 2008),meaning that cropping mixtures typically found in the study area cannot be simulated by SWAT in its present form (Kiniry et al., 2008).

Balancing the representation of such diverse vegetative covers in comprehensive tools like SWAT requires careful consideration of the objectives and the level of detail required to achieve a desired accuracy in simulating water fluxes and water quality. Incorporating accurate plant growth processes into hydrological models can improve the simulation performance and provide better decision aids. Robust models for crops, grasses and trees provide quantitative means to predict the hydrological consequences of various management decisions under different environmental and climatic conditions. These include harvesting schemes, replanting, fertilizer applications and control of undesirable plants. According to Kiniry et al. (2008), field-scale simulators like SWAT provide a general description of the growth of a vegetative canopy using deterministic relationships based on physiological or physical processes. Leaf growth is often represented by the leaf area index (LAI). Yield can be simulated using a harvest index (HI) approach, assuming yield is a fraction (the HI) of the total above-ground biomass (Fig.8-2). Such simulators can be readily applied to several plant types by deriving realistic plant parameters such as radiation use efficiency (RUE), maturity type, leaf angle through the light extinction coefficient and efficiency in partitioning the biomass through the HI.

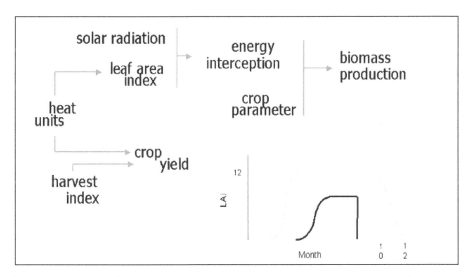

Figure 8-2: Parameterisation of the crop production in SWAT (Nietsch et al , 2002)

Deficiencies in adequately simulating the plant growth responses to water availability can fall under two categories: inadequate quantification of the process or the omission of a needed process in the simulator (McMaster et al., 2005). SWAT uses the Erosion Productivity Impact Calculator (EPIC) crop model concepts of phenological crop development, based on daily accumulated heat units, on a harvest index for partitioning grain yield, on the Monteith approach for potential biomass and on water and temperature stress adjustments. (Williams et al., 1989). A simple concept is used for simulating all the crops considered and SWAT is capable of simulating crop growth for both annual and perennial plants. Annual crops grow from planting date to harvest date or until the accumulated heat units equal the potential heat units for the crop. Perennial crops maintain their root systems throughout the year, although the plant may become dormant after frost (Arnold et al., 1998). The default number of heat units (HUSC) that control start and stop of growth are appropriate for an annual crop that is planted after danger of frost and allowed to dry-down before harvest, but not for many annuals. Reporting for a study in the Californian region, Tetratech (2004), noted that setting the initiation of growth at the default of 15% of the base zero heat units means that trees do not start leafing out until the beginning of April in many of the test watersheds, which is clearly too late. Also, setting the heat units for the "kill" operation to the default of 1.20 is probably too large for woody plants in arid areas, as it would mean that leaves are likely to be retained for a long period after the completion of the annual growth cycle.

The SWAT model simulates light (Photosynthetic Active Radiation; PAR) interception, assuming a constant value for the light extinction coefficient (k) of 0.6 for PAR. In the EPIC model, lower k values (e.g 0.4-0.6) are suggested for use in tropical areas (where the average sun angle is higher) and for wider row spacing (Williams et al., 1989). SWAT simulates the soil evaporation using the leaf area index (LAI). Using a moist soil surface and for a crop reaching a LAI of 5.8 (Kiniry et al., 2008) demonstrated that the LAI-based method predicts a total evaporation that is approximately 23% lower than the solar radiation interception-based method. They suggested an improvement in the radiation interception simulation, especially during the early growth because "the randomness in leaf distribution implicit in using the extinction coefficient approach is clearly violated in row crops at low LAI when there is minimum overlap among leaves, and second, because changes in canopy cover affect soil evaporation the most at low LAI".

The fraction covered by forested areas, their structure and species composition has a fundamental influence on the hydrological behaviour of a landscape (Wattenbach et al., 2005). The combination of the rooting strategy, interception losses, high surface roughness of trees, low albedo and great leaf areas lead to higher evapotranspiration rates than any other vegetation type under the same environmental conditions. The vegetation growth model in SWAT was originally developed for agricultural crops, containing the essentials to model the hydrological cycle for areas with annual crop production or perennial grass production (MacDonald et al., 2008). Dynamic forest growth is either not considered or is processed through simple parameterization. Reproduction and simulation of forest related processes including biomass accumulation, LAI development and root water uptake and related processes (transpiration and interception) is poor (Wattenbach et al., 2005, Johnson et al., 2009).

In SWAT, the default management operations for forest are 'Plant/begin growing season' and 'Harvest and kill', which means that trees are planted each year, then growth stops later in the year, the trees are all chopped down and converted to residue. Using these default settings in SWAT, the biomass values in forest are unrealistically low and drop to zero every year (Fig.8-3). The model assumes that the forest is growing from zero biomass at the beginning of each year. SWAT experiences problems in simulating biomass production for tropical and other climates even when all parameters in the crop database are correctly specified.

Cover specific modifications of the management (.mgt) files for each HRU are required to achieve seemingly reasonable biomass simulations. Consequently, due to the long maturation of forest cover (typical 35 years), the annual built up of biomass is very small. Further, since SWAT was developed for temperate regions, it is assumed by default that senescence and leave drop occurs every year: the vegetation goes into dormancy in the winter period.

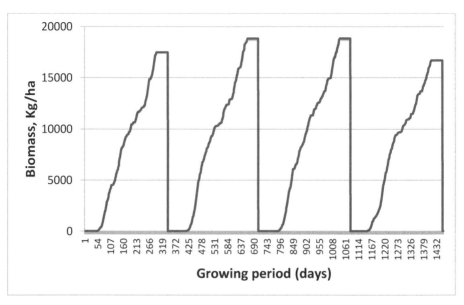

Figure 8-3: The evolution of biomass in a forest HRU in the SWAT default setting for the Upper Mara basin

8.3.2 The groundwater module

In SWAT, the representation of the aquifers does not allow the explicit simulation of underground water movements between subbasins (Notter 2010). The soil profile and shallow aquifer are physically disconnected, with the shallow aquifer being treated as a tank located somewhere below the soil profile (Fig. 8-4). There is active interactions between the soil profile and the shallow groundwater. Some of the interactions include: the wetting action on the soil profile due to changing water table, and the uptake of water by plant from the shallow aquifer. According to Kim et al. (2008), SWAT connects the soil profile and the shallow aquifer using two variables: the groundwater "revap" or groundwater evaporation (GW_REVAP) and the - threshold water depth in the shallow aquifer for "revap" to occur (REVAPMN). Revap is lost to the atmosphere and thus is not directly linked to the water content in the soil profile.

Table 8-1: Summary of the key elements in SWAT2005 (modified from Borah and Bera, 2003)

Model components/ capabilities	Hydrology, weather, sedimentation, soil temperature, crop growth, nutrients, pesticides, agricultural management, channel and reservoir routing, water transfer, with user interface and GIS platform
Temporal scale	Long term; daily or sub-daily steps.
Watershed representation	Sub-basins grouped based on climate, hydrologic response units (lumped areas with same cover, soil, slope and management), ponds, groundwater, and main channel.
Rainfall excess on overland	Daily water budget; precipitation, runoff, ET, percolation, and return flow from subsurface and groundwater flow.
Runoff on overland	Runoff volume using curve number and flow peak using modified Rational formula or Soil Conservation Service Technical Release 55 (SCS TR-55) method.
Subsurface flow	Lateral subsurface flow using kinematic storage model, and groundwater flow using empirical relations
Runoff in channel	Routing based on variable storage coefficient method and flow using Manning's equation adjusted for transmission losses, evaporation, diversions, and return flow
Flow in reservoir	Water balance and user-provided outflow (measured or targeted).
Overland sediment	Sediment yield based on Modified Universal Soil Loss Equation (MUSLE) expressed in terms of runoff volume, peak flow, and USLE factors.
Channel sediment	Bagnold's stream power concept for bed degradation and sediment transport, degradation adjusted with USLE soil erodibility and cover factors, and deposition based on particle fall velocity.
Reservoir sediment	Outflow using simple continuity based on volumes and concentrations of inflow, outflow, and storage.
Chemical simulation	Nitrate-N based on water volume and average concentration, runoff P based on partitioning factor, daily organic N and sediment adsorbed P losses using loading functions, crop N and P use from supply and demand, and pesticides based on plant leaf-area-index, application efficiency, wash off fraction, organic carbon adsorption coefficient, and exponential decay according to half lives.
BMP evaluation	Agricultural management: tillage, irrigation, fertilization, pesticide applications, and grazing.

129

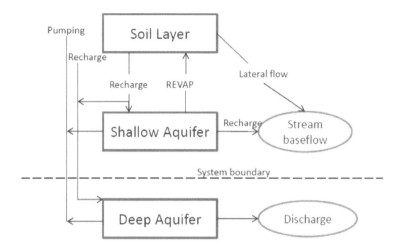

Figure 8-4: The partitioning of the groundwater in SWAT (modified from Vazquez-Amábile and Engel, 2005)

If no restrictions/forcing are applied on the model during calibration especially with the aquifer percolation coefficient (Rchrg_dp), and the threshold water level in shallow aquifer for baseflow (GWQMN) parameters, the model dumps water into the aquifers leading to high accumulation of water in the aquifer making it unavailable for the SWAT water balance. Notter (2010) recommended the provision for multiple separated deep aquifer systems that each comprise one or several subbasins, with groundwater discharge governed by parameters similar as for the shallow aquifer.

8.4 The adaptation of SWAT to tropical conditions

8.4.1 Adaptations of SWAT to conditions in Africa

In a SWAT modification named SWAT Water Balance (SWAT-WB), White et al. (2009) introduced a saturation deficit for each soil profile. This deficit replaces the SCS-CNII parameter for the calculation of the surface runoff and thus of the infiltration, whereby, the surface runoff is equal to the amount of rainfall minus the amount of water that can be stored in the soil before it saturates. According to the authors, "the SWAT-WB provided much more realistic spatial distribution of runoff producing areas. These results suggest that replacement of the curve number (CN) with a water balance routine in SWAT: significantly improves model predictions in monsoonal climates, provides equally acceptable levels of accuracy under more typical north American conditions, while at the same time greatly improving the ability to predict spatial distribution of runoff contributing areas". The model was tested on the Gumera watershed, a 1270 km^2, heavily (~95%) cultivated watershed located in the Blue Nile River Basin in Ethiopia.

Notter et al. (2009), working in the Pangani basin in Tanzania, introduced SWAT-PANGANI (SWAT-P) with several modifications especially on the auto-irrigation routine. The irrigation model in SWAT was changed to allow for interbasin transfer of excess water. According to Notter, 2010, a new input parameter "irr_eff" describing irrigation efficiency was introduced in SWAT-P. The values of "irr_eff" can range from 0 to 1, where 0 means none of the water abstracted from the source reaches the destination, and 1 means all water reaches the destination.

8.4.2 Adaptation of SWAT in other parts of the world

In the SWAT- DRAINMOD, Vazquez-Amábile and Engel (2005) modified the code in order to provide the soil water content layer by layer for every HRU. Using this information, it was possible to convert the soil moisture into a groundwater table level based on the relationship between the water table depth and the drainage volume, which is the effective air volume above the water table. SWAT has also been combined with the groundwater model MODFLOW (SWAT-MOD, Fig.8-5) to address the interconnectivity of the surface and groundwater phases (Sophocleous et al., 1999; Kim et al., 2008).

According to Kim et al., (2008) the groundwater component of SWAT does not consider distributed parameters, such as the hydraulic conductivity and the storage coefficient, due to its semi-distributed features. To resolve this challenge they swapped MODFLOW cells for SWAT HRUs, and used the HRU-Cell interface to allow for simulation of groundwater recharge and evapotranspiration in a distributed manner. For the simulation of the forest land cover, Watson et al., (2005) integrated physiological principles in SWAT for predicting the forest growth, based on 3-PG (SWAT-3PG). The 3-PG (Landsberg and Waring, 1997) is a dynamic, process based simulator of forest growth that predicts the net photosynthesis by forest stands on a monthly basis and allows process based calculations to estimate the forest growth in terms of a few variables. It is a generalized standalone simulator applicable to plantations and even-aged, relatively homogenous forests, based on (optimized) parameters that are related to tree physiology.

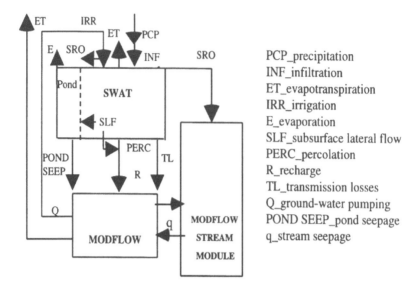

	PCP_precipitation
	INF_infiltration
	ET_evapotranspiration
	IRR_irrigation
	E_evaporation
	SLF_subsurface lateral flow
	PERC_percolation
	R_recharge
	TL_transmission losses
	Q_ground-water pumping
	POND SEEP_pond seepage
	q_stream seepage

Figure 8-5: Schematic block diagram of the SWAT/MODFLOW linkages (Sophocleous et al, 1998)

8.5 Conclusion

Despite the few shortcomings in the SWAT simulator, it was selected amongst the other available watershed scale models for this study. The model has widespread application in divergent environmental conditions. This extensive usage provides a good opportunity for benchmarking and comparing outputs with other studies under similar conditions. The development of the SWAT simulator is a continuation of modeling experience spanning a period of 30 years. The very large and vibrant SWAT user group ensures continuous sharing of information on the challenges and new improvements to tackle the shortcomings encountered. The simulator's open source code allows for manipulation and changing of the codes and reporting of the same to the developers and in the forum. The release of updated versions and documentations passes on the improvements to the wider user community leading to a more robust system. SWAT is also among the few basin scale hydrological models that can be used to simulate climate predictions and which is an important aspect in this study.

9. Hydrological modelling of the Mara basin using SWAT

9.1 Introduction

Many of the currently available distributed parameter simulators require watershed data in a geographic information system (GIS) format to facilitate model parameterization (Cotter et al., 2004). The AVSWATX version of SWAT2005 (Di Luzio et al., 2002) comprising of a preprocessor, interface and post processor under ArcView was used for the model setup. Mandatory GIS input files needed for the SWAT model include: the digital elevation model (DEM), land cover, and soil layers.

The first step in the SWAT modeling process is to delineate the sub-watersheds in the basin. This requires a DEM Map to define the network delineation using the threshold method. This is the threshold in area units flowing into a given point before it is designated as a stream. The lower the number the more streams and sub-basins will be created. The sub-basins are subsequently divided into HRUs by the user specified land use and soil percentage (Neitsch et al., 2002). According to Chaubey et al. (2005), the input DEM data resolution affects the SWAT model predictions by affecting the total area of the delineated watershed, the predicted stream network and the sub-basin classification. The result is a watershed that is broken down into several sub-basins. Each of these sub-basins drains to a particular point. That point can be chosen by the user or based on a pre-set resolution. The user defined outlets for the sub-basins are selected based on source points, calibration points, elevation bands and/or land use. The subwatersheds are then further subdivided into HRUs that consist of homogeneous land use, management, slope and soil characteristics. Multiple HRUs are created by specifying sensitivities for the land use and soil data that will be used to determine the number and kind of HRUs in each watershed.

Weather data from stations within the region were incorporated to provide the most representative meteorological data data available. Weather data required by SWAT include: precipitation, temperature, solar radiation, wind speed, and relative humidity. In the absence of observed climatic data the weather input is estimated using the SWAT weather generator. Measured flow and water quality data specific for the watershed is identified to perform calibration and validation of the SWAT model within the watershed. The SWAT model is calibrated for flow, at daily, monthly and annual time scales.

A description of the overall modeling process: data preparation, model set up, model calibration and validation for the Upper Mara is presented the following sections. The input data for this study was sourced locally from the national depositories or from global databases. While some datasets, like the SRTM DEM, required little adjustments, others had to be refined through application of additional data processing techniques. SWAT model inputs requiring expert manipulation include: the land use maps and the climatic data inputs. Land cover map were prepared using satellite imagery through the generations of thematic maps (§5.2). Geostatistical tools were applied on the climatic data to prepare interpolated time series. All key input databases including the DEM, land use map and soils maps were projected to the same spatial reference (Annex 7).

9.2 Data preparation

9.2.1 Topographic data

The Shuttle Radar Topography Mission (SRTM) 90m X 90m DEM was used. The characteristics of the DEM used in this study have been described in section §4.2.2.1 The elevation in the study area is composed of clear defined elevation bands which cut across the basin in a north-west to south-east direction, while the slope flows from Northeast to south-west direction (Fig 9-1).

9.2.2 The soil map

The KenSOTER database used for the soil information has been described in §4.2.2.2 The KENSOTER database was selected because the global SOTER system is expected to replace the FAO-UNESCO Soil Map of the World (SMW). The SMW was the first internationally accepted inventory of world soil resources. The soil hydraulic conductivity (Ksat), a parameter required for the SWAT soil database is not provided in the KenSOTER database. The K_{sat} was calculated using a pedo-transfer function developed by Jabro (1992) and modified by (Droogers et al., 2007). The K_{sat} was calculated using the following equation,

$$K_{sat} = \exp\left(11.86 - 0.81\log(silt) - 1.09\log(clay) - 4.64(Bulk\ density)\ mm/hr\right.$$

--------------------9-
1

134

Where: K_{sat} is saturated hydraulic conductivity (mm/hr),

Silt is silt content (%)

Clay is clay content (%)

9.2.3 The land use map

The existing FAO AFRICOVER land cover map (FAO, 2005) was used as the baseline map. The AFRICOVER-Kenya Map is a spatially aggregated multipurpose land cover database produced from visual interpretation of digitally enhanced LANDSAT TM images (Bands 4, 3, 2) acquired mainly in the year 1999 (FAO, 2005). The land cover class, according to the original AFRICOVER dataset is a broad group and consists of several descriptive classes. The AFRICOVER classes were reclassified into SWAT usable land use classes in the SWAT look-up table. The description of the land cover classes from FAO and their SWAT equivalents are given in Table 9.1 below. The GIS input maps for the DEM, soil and land cover are given in Fig 9.1

9.2.4 The hydroclimatic data

The Kenya Meteorological Department (KMD) is the official custodian of climatic data in Kenya. The KMD data was complemented with privately collected data and with global climatic sources, including the CRU2.0 database of the University of East Anglia (New et al., 2002). The rainfall stations have data series of varying lengths and different levels of missing data. Two methods, the nearest neighbours (NN) and the inverse distance weighting (IDW) methods were used to impute the missing data in the databases and to reduce the effect of the topography. According to Cho et al., (2009), as the number of rain gauges used in the simulation decreases, the uncertainty in the hydrologic and water quality model output increases exponentially.

In SWAT, the weather station nearest to the centroid of each sub-basin is taken as the location for the precipitation to be used in the simulation. Schuol and Abbaspour (2006) noted that unrealistic weather data are generated by SWAT if a weather station is assigned to a subbasin that has only a few measured values or many erroneous values. According, to Grimes and Pardo-Igúzquiza (2010) the benefits of geostatistical analysis for rainfall include the ease of estimating areal averages, the estimation of uncertainties and the possibility of using secondary information like topography. In the nearest neighbour method, the rainfall stations closest to the stations with missing data were used to fill in the gaps. By using the neighbourhood stations any missing data in one station was filled in from the other stations. The principle of shared similarities of the stations to each other due to close spatial proximity is assumed.

In the IDW method, weights for each sample are inversely proportionate to its distance from the point being estimated. The method was used to develop a time series of rainfall data. All the stations with long term data were used in the algorithm to determine the weighted rainfall for the new station. Missing portions of any station data were filled with this series data.

$$P_x = \frac{\sum_{i=1}^{N} \frac{1}{d^2} P_i}{\sum_{i=1}^{N} \frac{1}{d^2}} \quad \text{---} 9.2$$

Where, P_x = estimate of the average basin rainfall, p_i = rainfall values of rain gauge i, d_i = distance from gauge I to the centroid of the basin, N = number of gauges.

The data series of the 1970-1977 periods provided the most complete data series and was used for this study. The CRU2.0 (New et al., 2002) long term (1961-1990) climatic data, on precipitation, temperature, evapo-transpiration and number of wet days for the centroid was compared with that of two WMO climatic stations (namely, Kericho and Narok) lying to the north west and south east axis of the basin. CRU2 data is on a 1/2 degree grid, with a time range 1901-2006. The variables are precipitation (mm/month), mean minimum temperature and mean maximum temperature per month. Each of these variables is has cell coordinates (0, 0) in the lower left, increasing to the right (New et al., 2002) The characteristics of a virtual station centroid to the basin were found to lie somehow between those of the Kericho and Narok stations for rainfall, number of wet days and temperature, (Fig.9-2). The use of the mean of the two stations removed biases towards any of the stations. The generated centroid thus had characteristics mid-way between those of Kericho and Narok. The average records from the two stations were used to calculate the parameters in the weather generator (.wgn) file. The .wgn file was created using the WGNmaker4.xlsm tool (Boisrame, 2011) an excel macro designed to calculate the weather statistics needed to create user weather station files for SWAT. The inputs to the WGNmaker4.xlsm include daily datasets for precipitation (mm), temperature (max. and min, $^\circ$C), solar radiation (MJ/m^2/day), and wind speed (m/s).

Table 9-1: Description of the land cover classes used in SWAT

Standard Description	LCC Label	Common name	SWAT code
		herbacious	
Rainfed Herbaceous crop, Small Fields	HR4	crop	RFHC
Continuous Rainfed Small fields [cereal]	Hm4	agric closely	AGRG
Rainfed Shrub Crop, Large Fields - Tea	SL47V-t	tea plantation	RFTT
Closed to very open herbaceous with sparse trees and sparse shrubs	2H(CP)78	herbs	HERB
Open woody with closed to open herbaceous	2WP6	upland maize	MAIZ
Rainfed Herbaceous Crop, Large to Medium Fields	HD4	lowland maize	CORN
Closed to very open herbaceous with sparse shrubs	2H(CP)8	grassland	PAST
Open trees (broadleaved deciduous) with closed to open shrubs	2TO28	open forest	FRST
Trees Plantation – Large Fields, Rainfed Permanent	TL47PL	tree plantation	FRST
Closed trees with closed to open shrubs	2TC8	natural forest	FRST
Permanently cropped area with small sized field(s) of rainfed shrub crop	SR47V	shrub crop	RFSC
Very open shrubs with closed to open herbaceous and sparse trees	2SVJ67	shrubs	SHRB
Open general shrubs	2SOJ67	shrubs	SHRB
Rainfed Herbaceous Crop, Large Fields - Wheat	HL4-w	wheat	RFWC
Rainfed Herbaceous Crop, Medium Fields - Wheat	HM4-w	wheat	RFWC
Rainfed Herbaceous Crop,	HL4	agric generic	AGRL

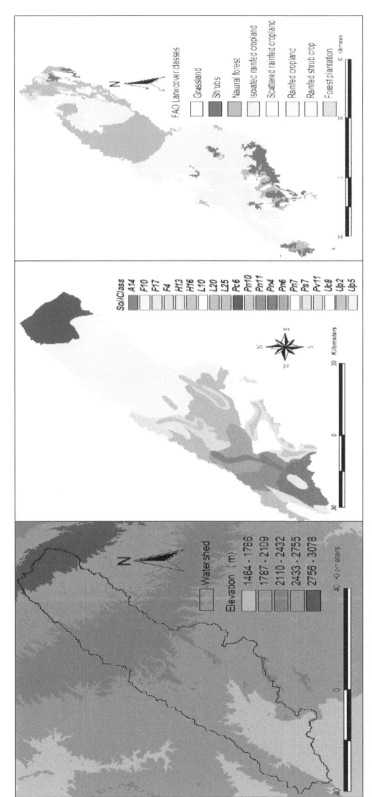

Figure 9-1: Map inputs for the baseline SWAT model

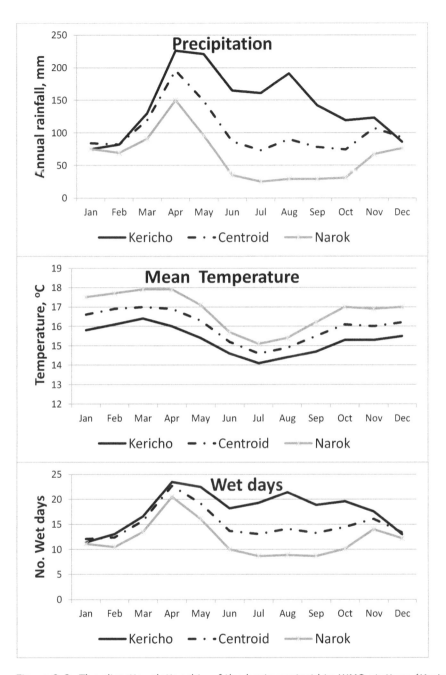

Figure 9-2: The climatic relationship of the basin centroid to WMO stations (Kericho, Narok) based on CRU 2.0 data (New et al., 2002)

For the precipitation input, the station created using the NN or IDW procedures in each subbasin was used as model input. For the other climatic parameters inputs including min. and max. temperatures, the data from the basin's centroid was used.

9.2.5 The stream flow data

Stream flow data is available from three stations 1LA02, 1LB03 and 1LA04. The 1LA02 and 1LB03 stations are located at the midsection of the watershed, while the 1LA04 station is located on the downstream most edge of the study area. The characteristic of the flow has been described in section §4.2.1.2

9.3 The model build-up

9.3.1 DEM processing and sub-basin delineation

In the lower flat plains of the watershed and due to the coarse resolution of the DEM, the delineation of the catchment was problematic. To correct the anomaly, a shapefile of digitized streams from World wildlife fund- Hydrological data and maps based on SHuttle Elevation Derivatives at multiple Scales (WWF- HydroSHEDS Lehner, (2005) was used to burn these into the DEM, Fig 9-3.

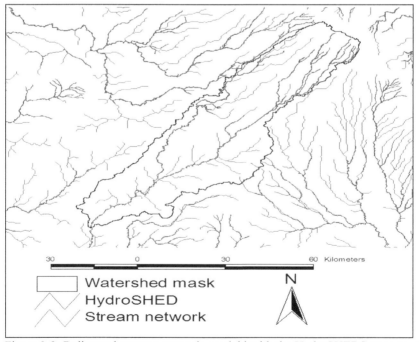

Figure 9-3: Delineated streams network overlaid with the HydroSHEDS streams map

The watershed was divided into 3 sub-basins. The lowest points for the each sub-basin are located at the site of the stream gauging stations 1LA03 1LB02 and 1LA04. Point 1LA04 is located downstream of the confluence of the Amala and Nyangores rivers. Point 1LA04 is the south most point in the study area and marks the end of the human activities including cultivated agriculture that may have direct impact on the water resources, and the start of the Maasai Mara natural reserve.

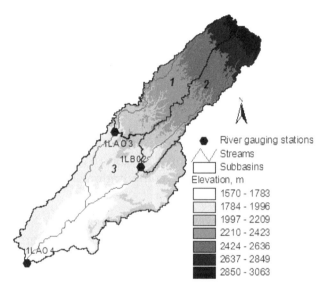

Figure 9-4: The delineation of the watershed into sub-basins

9.3.2 Watershed decomposition

The SWAT model decomposes a watershed to sub-basins and Hydrologic response units (HRUs). HRUs represent the unique combinations of soil and land cover within each sub-basin at a specified distribution and are considered to be hydrologically homogeneous. HRU distributions are set by the user as a percentage value for soil and land cover. The distribution percent is used by the model so that only combinations of soil and land cover that are greater than the set distribution percentage are considered. Soil and land cover combinations below the set HRU distribution percentage are lumped into the combinations above the set distribution percentage and therefore not represented in the model. HRUs provide the greatest resolution for parametrization in SWAT; however, HRUs do not possess spatial orientation with respect to each other within a sub-basin (Nietsch et al., 2002). For this study all the 19 soil (§9.1.2) and 15 land use (§9.1.3) classes were used in the overlaid map with 0% threshold for both land use and soil resulting into 63 HRUs. The number of HRUs in the landuse classes differs from one class to another.

9.4 The model sensitivity analysis

A sensitivity analysis is usually the first step towards model calibration and is performed to identify the parameters that have the greatest influence on the model results. According to Cho and Lee (2001), a sensitivity analysis seeks to answers several questions including: (a) where data collection efforts should focus; (b) what degree of care should be taken for parameter estimation and (c) the relative importance of various parameters. The parameters in SWAT vary by sub-basin, land use, or soil type, hence increasing the scale in the discretization (or threshold area) increases the number of parameters substantially. While some of these parameters represent measurable quantities and hence can be estimated directly from field data (or from literature), other parameters are empirical or SWAT-specific. A sensitivity analysis method should be both computationally efficient and robust.

A parameter sensitivity analysis was performed using the built in sensitivity analysis tool of AVSWATx which uses the Latin Hypercube – One At a Time (LH-OAT) method (Van Griensven et al., 2006). The LH-OAT method performs LH sampling followed by OAT sampling. A total of eighteen parameters with a sensitivity index > 0 were identified as sensitive with regard to flow (Table 9-2). According to Van Griensven et al., (2006) parameter with a global rank 1 is categorized as 'very important', rank 2–6 as 'important', rank 7–20 as 'slightly important' and rank 28 as 'not important'. The Soil Conservation Service (SCS) curve number (CNII) was identified as the most sensitive and hence "very important" parameter to stream flow for this watershed. The curve number indicates the runoff potential of an area for the combination of land use characteristics and soil type. Higher curve numbers translate into greater runoff. Curve numbers are a function of hydrologic soil group, vegetation, land use, cultivation practice, and antecedent moisture conditions. The CNII parameter is of primary influence on the amount of runoff generated from a hydrologic response unit and hence a relatively large sensitivity index was expected. The parameter which depends on the percentage of imperviousness in the land cover type and the soil group is important especially in the study area with forest and cultivated land as the major cover groups and little urban settlement influence.

The other "important" parameters, were the groundwater recharge to deep aquifer (rchrg_dp), the threshold depth of water in the shallow aquifer required for return flow to occur (GWQMN), the available water capacity (SOL_AWC) and groundwater "delay" coefficient (GW_DELAY). The rchrg_dp controls the fraction of the percolated water that will flow to the deep aquifer. A high rchrg_dp value (near 1) indicates more allocation of percolated water to the deep aquifer. In SWAT the water that percolates through the unsaturated zone is immediately divided between the shallow and deep aquifers. The deep aquifer fraction will not produce any runoff in the basin and is thus water that is lost to the basin.

The GWQMN regulates the water accumulation in the aquifer. The groundwater flow to the reach is only allowed if the depth of the water in the shallow aquifer is equal or greater than the GWQMN. The GWQMN parameter has wide threshold range (0-5000mm). SOL_AWC is the difference between the field capacity and the wilting point. The parameter is dependent on the soil textural class and ranges from 0.02 to 0.3 with typical values of 0.04, 0.24 and 0.21 for sand, loam and clay respectively. GW_DELAY is the lag between time the water exits the soil profile and enters shallow aquifer, and depends on the depth of the water tble and the hydraulic properties of the geological formation.

9.5 The model calibration

Multi-gage observed data if available may be used for implementing a distributed approach to calibration, where observed and simulated outputs are compared at multiple points on a watershed. Moreover, it may be advantageous to employ a multi-step approach to auto-calibration (Van Liew and Veith, 2009).

The period between 1970 and 1977 provided data with comparatively minimal missing data points and was used in this study for the three gauging stations. 1970 was used for model warm-up, 1971-1974 for model calibration and 1975-1977 for model validation.Manual calibration was attempted but discarded due to two fundamental challenges: manual calibration of the large number of HRUs model was labor intensive and the large number of non unique solutions was not only confusing, but produced no better results than the default simulation.

Table 9-2: Ranking of the parameters sensitive to flow in the SWAT modelling

Rank	Parameter	Lower limit	Upper limit	Parameter description	Msi*
1	CN2	-50	50	SCS runoff curve number II	2.59
2	rchrg_dp	0	1	Groundwater recharge to deep aquifer	1
3	GWQMN	0	5000	Threshold depth of water in the shallow aquifer required for return flow to occur (mm)	0.87
4	SOL_AWC	-50	50	Available water capacity (mm/mm soil)	0.5
5	GW_DELAY	0	100	Groundwater delay (days)	0.33
6	ALPHA_BF	0	1	Baseflow alpha factor (days)	0.3
7	SLOPE	-50	50	Average slope steepness (m/m)	0.2
8	ESCO	0	1	Soil evaporation compensation factor	0.19
9	Sol_k	-50	50	Soil conductivity (mm/h)	0.17
10	Sol_z	-50	50	Soil depth	0.14
11	Canmx	0	10	Maximum canopy index Runoff	0.08
12	CH_K2	0	150	Effective hydraulic conductivity in main channel alluvium (mm/hr)	0.06
13	Sol_alb	0	1	Moist soil albedo	0.04
14	surlag	0	10	Surface runoff lag coefficient	0.02
15	EPCO	150	50	Plant evaporation compensation factor	0.02
16	BLAI	-50	50	Maximum leaf area index	0.02
17	GW_REVAP	0.02	0.2	Groundwater "revap" coefficient	0.01
18	REVAPMN	0	500	Threshold depth of water in the shallow aquifer for "revap" to occur (mm)	0.01

*msi= mean sensitivity index

The process of manual calibration requires a high degree of expert knowledge of the model and the system and is characterized by subjectivity in the strategy employed to adjust the parameter values, as well as in the criteria (mainly visual) used to judge the goodness-of-fit of the model simulation (Blasone et al., 2007). Also, due to the large amount of missing data for the period under consideration, the base flow separated with the available filters was only estimation.

According to Arnold et al., (1995b) some type of baseflow filter is used to provide an average annual ratio of baseflow to surface runoff for the calibration of model runs. Baseflow separation was therefore performed for 1971 and 1974 to give an indication of the baseflow fraction. Three different digital filters were used to perform the base flow separation for 1971 and 1974. Although digital filter method has no physical meaning, it removes the subjective aspect from manual separation, and it is fast, consistent, and reproducible (Arnold et al., 1995b).

The Baseflow Program (Bflow, Arnold et al. 1995b) makes three passes (forward, backward and forward) with each pass resulting in less base flow as percentage of the total flow. The authors recommend the use of first pass as default in the absence of site-conditions, although it overestimates the baseflow compared to manual separation techniques, Arnold et al. (1995b), and Arnold et al. (1999) found measured values to fall in the midpoint of first and second passes.

The Water Engineering Time Series PROcessing tool (Wetspro, Willems, 2009) can be used to conduct the; sub flow filtering, peak flow selection and related hydrograph separation for quick flow and slow flow periods, and related low flow selection (Willems 2009).

The WHAT -Web-based Hydrograph Analysis Tool, Lim et al., (2005) has three base flow separation modules, the local minimum method and two digital (one parameter digital filter method - same algorithm used in BFLOW filter , and two parameter digital filter method- filter parameter and BFImax) filter methods, are available in the WHAT system. The mean base flow fraction of flow from the analysis of the three methods (Bflow, Wetspro, and WHAT)and adopted for this study was found to lie between 60-70%.

The SWAT Calibration and Uncertainty Programs (SWAT-CUP) stand-alone program was used for the calibration and validation of the SWAT model. The program links the Generalized Likelihood Uncertainty Estimation method (GLUE; Beven and Binley, 1992), the Parameter Solution method (ParaSol; Van Griensven and Meixner, 2006), the Sequential Uncertainty Fitting (SUFI2; Abbaspour et al., 2004), the Markov chain Monte Carlo method (MCMC; Kuczera and Parent, 1998) and the Particle Swarm Optimization method (PSO; Kennedy and Eberhart, 1995). It enables sensitivity analysis, calibration, validation and uncertainty analysis of a SWAT model.

The autocalibration was performed using the SUFI-2 and the ParaSol algorithms. In both ParaSol and SUFI2, the parameters can be changed in three ways; a - an absolute change of value, v – replacement of the value and r – a relative change of value. The choice of the change method was done as per the protocol set out in the "changepar" section of the sensitivity, autocalibration and uncertainity analysis manual of (Van Griensven 2006). Two "slightly important" parameters; the Sol_z and SLOPE (rank 7 and 10 respectively) were not included in the autocalibration. Sol_z caused the model to crash (a problem also repeatedly reported elsewhere in the SWAT model forums), the slope is a basin parameter and related to the DEM. The two last parameters with sensitivity indices < 0.02 were also not used in the auto calibration routine.

For the autocalibration, stream flow data from 1LA03, 1LB02 AND 1LA04 stations was used. The parameter ranges are set through available data, literature and suggestions from the SWAT user manual. During autocalibration, most of the parameters were set in the ranges recommended in the SWAT calibration manuals (Van Griensven et al., 2006; Abbaspour et al., 2004). Only two parameters, the GWQMN and the Rchrg_dp were constrained in order to limit the amount of water that is routinely percolated to the aquifers. Through preliminary trial and error runs, the parameters GWQMN and Rchrg_dp were set to allow maximum value change up to 200 mm and 0.1 respectively.

After the comparison of the initial results of the autocalibration, there was no significant difference in the NSE from ParaSol and SUFI algorithms. The ParaSol method was therefore selected for subsequent autocalibration procedures. The ParaSol algorithm is faster and required less interactive attention, as the iterations were completely automated unlike the SUFI-2 where the number of runs per iterations is limited to 500 runs, and the modeller has to set new parameter limits for subsequent iterations (Abbaspour et al., 2004). The ParaSol incorporated in AVSWATx was therefore adopted for this study. The ParaSol method calculates the sum of the squares of the residuals (SSQ) as the objective function (OF). The OF is based on matching a simulated model output to observation (measured) time series. The ParaSol aggregates these objective functions into a global optimization criterion (GOC), minimizes the OF or GOC using the Shuffled Complex Evolution Uncertainty Analysis (SCE-UA) algorithm and performs an uncertainty analysis with a choice between two statistical concepts (Abbaspour, 2004). The shuffled-complex-evolution algorithm is slower to optimize than the parameter estimation (PEST, Doherty, 2005)) optimization package, but makes no assumptions about the shape of the objective function, and is therefore less likely to be trapped in local minima (Marshall, 2005).

The acceptability and usability of a model to simulate or predict physical processes depend on how well it models compared to observed data.

Hydrological models can be assessed by their goodness of fit to statistical measures - based on an objective function - and by comparison to the water mass balance in the watershed. Both metric and non metric performance measures are used in this study. The goodness-of-fit measures used include the Nash-Sutcliffe efficiency (NSE) value Nash and Sutcliffe, (1970), the percent bias (PBIAS) and the Root Mean Square Error-observations standard deviation ratio (RSR). The three methods are based on three parameters (Y_i^{obs}, Y_i^{sim}, and Y_i^{mean}). Y_i^{obs} is the ith observation for the constituent being evaluated, Y_i^{sim} is the ith simulated value for the constituent being evaluated; Y^{mean} is the mean of observed data for the constituent being evaluated and n is the total number of observations.

The Nash-Sutcliffe efficiency (NSE) is a normalized statistic that determines the relative magnitude of the residual variance as compared to the measured data variance (Nash and Sutcliffe, 1970). The relationship between NSE and SSQ is

$$SSQ = \Sigma_{i=1}^n \left(Y_i^{obs} - Y_i^{sim}\right)^2 \text{------------------------------9.3a}$$

$$NSE = 1 - \left[\frac{1}{\Sigma_{i=1}^n (Y_i^{obs} - Y_i^{mean})^2} \cdot SSQ\right] = 1 - \left[\frac{\Sigma_{i=1}^n \left(Y_i^{obs} - Y_i^{sim}\right)^2}{\Sigma_{i=1}^n (Y_i^{obs} - Y_i^{mean})^2}\right] \text{--------9.3b}$$

Where: NSE ranges between $-\infty$ and 1 (1 inclusive), with NSE=1 being the optimal value. Values between 0.0 and 1.0 are generally viewed as acceptable levels of performance, whereas values <0 indicates that the mean observed value is a better predictor than the simulated value, which indicates unacceptable performance.

The NSE is recommended for use by the American Society of Civil Engineers (ASCE) (Moriasi et al., 2007). Due to its widespread usage, it provides extensive information on reported values. It was found by Sevat and Dezetter (1991) to be the best objective function for reflecting the overall fit of a hydrograph. An NSE of 0.50 - 0.65 was set as satisfactory by Moriasi et al. (2007). The PBIAS measures the average tendency of the simulated data to be larger or smaller than their observed counterparts.

$$PBIAS = \left[\frac{\Sigma_{i=1}^n \left(Y_i^{obs} - Y_i^{sim}\right) * 100}{\Sigma_{i=1}^n Y_i^{obs}}\right] \text{-------------------------9.4}$$

The optimal value of PBIAS is 0, with low-magnitude values indicating an accurate model simulation. Positive values indicate model underestimation bias, and negative values indicate model overestimation bias.

The RSR is calculated as the ratio of the RMSE and the standard deviation of the measured data. RSR incorporates the benefits of the error index statistics and includes a scaling/normalization factor, so that the resulting statistic and reported values can apply to various constituents (Moriasi et al., 2007).

$$RSR = \frac{RMSE}{STDEV_{obs}} = \frac{\left[\sqrt{\sum_{i=1}^{n}(Y_i^{obs} - Y_i^{sim})^2}\right]}{\left[\sqrt{\sum_{i=1}^{n}(Y_i^{obs} - Y_i^{mean})^2}\right]} -----------9.5$$

RSR varies from the optimal value of 0, which indicates zero RMSE or residual variation and therefore perfect model simulation, to a large positive value. The lower RMSE, the lower the RSR, and therefore, the better the model simulation performance

A graphical comparison between simulated and observed hydrographs should always be undertaken in any study involving computed and simulated hydrograph comparisons (Green and Stephenson, 1986). Despite its obvious shortcomings of subjectivity (and hence irreproducibility) and inapplicability in large data sets, visual inspection is a "powerful expert system for simultaneous, case specific multi-criteria evaluation which provides results in close accordance with the user's needs" (Ehret and Zehe, 2010).

9.6 SWAT model performance and improvements

The NSE for the calibrated simulations for the three stations is given in Table 9-3. Since the objective of using the SWAT model was to adapt a hydrological model for the prediction of processes taking place in the watershed. The models should therefore, as much as possible, meet some minimum statistical performance criteria, as well as reflect the physical processes in the catchment. The threshold set for this study was NSE>=0.5 for monthly time step, which is consistent with the criteria spelt out in Moriasi et al. (2007).

Table 9-3: Model statistical performance for multi-stage calibration stations

River	stations	NSE	% deviation from observed flow
Mara	1LA04	0.29	-30
Nyangores	1LA03	0.13	24
Amala	1LB02	-5	36

Hydrological models can be assessed by their goodness of fit to an observed flow series, based on an objective function, and by an analysis of the water balance in the watershed, as compared to observations. From Table 9-3, the results of the calibration were not satisfactory, even after manipulation of the parameter ranges for the model. A change in the model set up was therefore necessary. The SWAT model was systematically improved in order to make its performance agree to the set minimum modelling performance standards and to mirror the hydrological processes in the watershed.

In order to get the SWAT model to satisfactory level, several improvements were made on the default SWAT model that was developed using the FAO-AFRICOVER land cover map of 1999 and the precipitation data for individual stations (§ 9.3). Whereas the Nash-Sutcliffe efficiency (NSE) was the index used for model statistical performance, the total water yield was the indicator for water balance performance. Improvement changes were made with and without recalibration of the model. Table 9-4 summarizes the changes made on the model inputs and management files throughout the improvement process. The improvement changes were first performed to improve the NSE and later the water balance, and are described in details in the following sections.

Table 9-4: Summary of changes on the SWAT model

ID	Improvement target	Land map	Precipitation	Management files
		Files changed		
1*	Objective function	AFRICOVER	Individual stations	default setting
2	Objective function	Landsat 1976	Individual stations	default setting
3	Objective function	Landsat 1976	Interpolated	default setting
4	Water balance	Landsat 1976	Interpolated	Perenial HRUs
5	Water balance	NDVI	Interpolated	Perenial HRUs
6	Water balance	NDVI	Interpolated	Perenial + Crops
7	ET forest	NDVI	Interpolated	Cropdat changed

*Default simulation

9.6.1 The improvement of the objective function.

The target for these improvement efforts was to have the SWAT model performance indices attain the performance criteria for hydrological models set out in Moriasi et al. (2007). Since rainfall and stream flow data for the period 1970-1977 was used for the model calibration and validation, and in cognizance of land cover change dynamics in the study area, the use of a latter day land cover map (the FAO 1999 map) was assessed as a potential source of error. The hydrology component of the SWAT model uses the curve method. The CNII is the most sensitive parameter, and is a function of the soil group and land cover. Therefore the representation of the wrong land cover type also affects the parameter sensitivity and response. A land cover map was created using Landsat MSS data to synchronise the data sets period under consideration. The FAO land map was substituted with the new land cover map and all the other model inputs and parameterization kept unchanged (Fig 9-5). Only the ILA04 station was used in this improvement process and the model improved from NSE of 0.29 to a new NSE of 0.43.

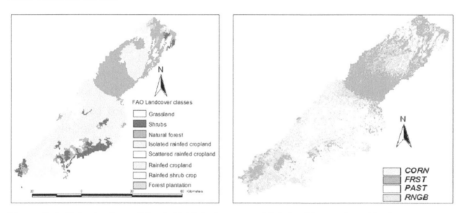

Figure 9-5: Substitution of the AFRICOVER map (right) with Landsat 1976 map

The SWAT model employs a simplified system for climatic data input. The station located nearest to the subbasin's centroid is used as the input for that subbasin. While this maybe applicable in areas with homogeneous terrain, it maybe problematic in regions where spatial heterogeneity is high or where data are sparse. Due to the geospatial heterogeneity of the study area, with elevation changing from 1800 to 3000m, the use of data of a single station for the basin or sub-basin was assessed as inappropriate.

A simple cluster and average (CA) method was implemented in this study. The method involved the clustering of the rainfall stations in a scatterplot. The average annual rainfall (mm) is plotted against the elevation (m). The relation betweeen the topographic elevation and the mean long term annual precipitation for the rainfall stations was a polynomial fit. Three clusters were distinguised, Cluster 1; Low altitude-low rainfall, cluster 2; Medium altitudes – high rainfall, and cluster 3; High altitudes – medium rainfall. Figure 3 represents the clustering of eleven (11) stations based on elevation and rainfall. The selection of the clusters was performed qualitatively by visible subject analysis of the plots. Due to their indiscreet positioning some stations were dismissed as noise and not clustered. Using the clustering principle where "patterns within a valid cluster are more similar to each other than they are to a pattern belonging to a different cluster" (Jain et al., 1999), three cluster classes were qualitatively established. Once the clusters were identified, the rainfall stations within the clusters were used to generate new, virtual stations by averaging stations' daily precipitation in each cluster (Table 9-5).

The clustering system was validated graphically by use of the areal interpolation of the annual rainfall using the ordinary Kriging technique. Using the ordinary kriging technique with a gaussian transformation, all the point rainfall stations were interpolated to a spatial distribution of rainfall over the study area. After re-classification, three zones were produced The zonings generated by the kriging interpolation were consistent with the clusters produced with the simple scatterplot clustering. The spatial interpolation indicates that the majority of the area upstream of the midsection experience higher than average rainfall, downstream had lower rainfall..The clusters identified represent a low altitude-low rainfall zone, a medium altitude-high rainfall zone and a high altitude-medium rainfall zone (Fig. 9-6). Three synthetic rainfall stations developed in these clusters and the new stations used as the precipitation inputs to the model.

The calibrated model performance using observed data for 1LA04 improved to NSE= 0.58. The NSE for the other two stations was considerably low (0.38 and 0.26 for Amala and Nyangores respectively).the graphical representation of the multi-stage calibration processs for 1970-1974 is shown in Fig 9-7. With the satisfactorily acceptable results from 1LA04, single station calibration using only 1LA04 was performed for the next improvements. The unsatisfactory performance for the 1LA03 and 1LB02, especially with the large amount of missing data (1LB02) and gross underestimation (1lA03) made it practically impossibble to rely on these stations for the SWAT model calibration. The poor performance of SWAT at the Nyangores station has also been experienced by other researchers. Mango et al. (2012)

obtained a NSE=0.085 and attributed the poor model performance to using data from limited number of raingauges/the very coarse spatial distribution of climate stations in the catchment.

Table 9-5: Stations used to generate new, virtual, precipitation stations with the Cluster and Average (CA) method

ID	LAT	LONG	Av. Rfall, mm	Elev. (m)	New station	Description of new stations
903507 9	-0.75	35.37	1465	2012	Pcp_med d	Medium altitude high rainfall
903526 0	-0.82	35.35	1627	1916		
903522 7	-0.78	35.33	1332	1951		
903526 5	-0.78	35.35	1343	1951		
903524 1	-0.42	35.73	1118	2865	Pcp_high	High altitude medium rainfall
903522 8	-0.45	35.8	1263	2957		
903532 4	-0.48	35.63	980	2650		
913500 8	-1.0	35.23	956	1646		
913501 9	-1.1	35.38	788	1829	Pcp_loww	Low altitude low rainfall
913501 0	-1.2	35.25	812	1826		
903533 4	-0.8	35.6	750	1942		

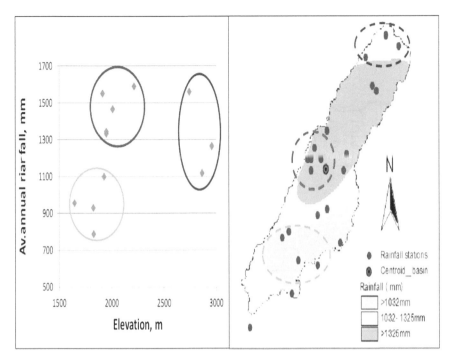

Figure 9-6: Scatter plot of the rainfall stations (left) and the corresponding spatial extent of the interpolation zones in the study area

9.6.2 The improvement of the water balance

While the NSE of the SWAT model was satisfactory, the other indicators of the model performance were not yet within acceptable ranges. The water yield was 16% lower than the observed flow. Arnold et al., (2012) recommended a deviation of +-10% for yields to be considered satisfactory. Default information entered in the management files in SWAT (.mgt) by the simulator interface creates problems for realistic simulation of landcover types specific for an area.

SWAT develops the biomass simulation primarily on the basis of the information in the crop database, which maybe modified to reflect appropriate parameters for regional cover types. To better reflect the catchment processes, the initial LAI, initial biomass and the heat units were adjusted according to Table 9-6. Since the model was simulated in a catchment in which permanent land cover types were growing, the Initial Plant Growth Parameters (IGRO) was changed.

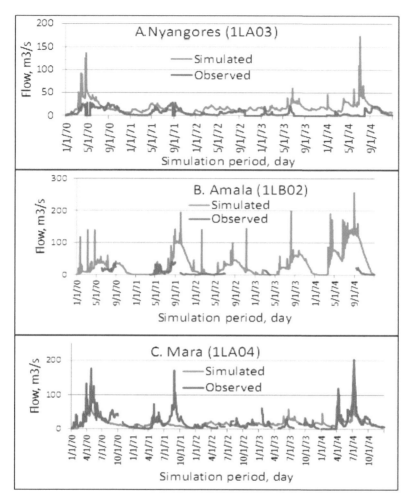

Figure 9-7. Time series for simulated and observed flows at three gauging stations on the Mara river

IGRO is the land cover status code, which informs the model whether or not a land cover is growing at the beginning of the simulation, (0 for no land cover growing, 1 for land cover growing). Plant_ID is the land cover identification number, and is the numeric code for the land cover given in the plant growth database. LAI_INT is the initial leaf area index, BIO_INT is the initial dry weight biomass (kg/Ha).

According to the SWAT documentation (Neitsch et al., 2002) trees go dormant as the day length nears the shortest or minimum day length for the year. During dormancy no growth takes place. Also once trees enter dormancy the tree leaf biomass is converted to residue and the LAI for the tree species is set to the minimum value allowed ('Harvest and Kill'). This has been changed to allow for an unlimited growth and biomass accumulation in the forest, taking into account the natural senescence processes. The initial biomass in the natural evergreen land use classes including forest and shrubs were initialized with values obtained in the literature to simulate mature tropical plants, Asner et al., (2003), Breuer et al., (2003), Houghton, (2005).

The potential heat unit (PHU LT) is the total number of heat units required to bring a plant to maturity, and is calculated using 1) long-term maximum and minimum temperature data, 2) the base or minimum temperature required by the plant for growth, 3) and the average number of days for the plant to reach maturity. For tree species, the PHU parameter is an estimate of the growing season, i.e., the amount of time between budding and leaf senescence. Once vegetation reaches maturity in the model, the leaf area index for the plant is set to zero and the vegetation no longer intercepts rainwater nor takes up water from the soil for the purpose of growth. Therefore, after plant maturity, evapotranspiration due to the vegetation does not occur. This modeling approach has a significant effect on soil moisture content and water yield for land covers with vegetation that has reached maturity (von Stackelberg et al., 2007).

Since the permanent cover types in the study area do not stop transpiring and certainly continue to intercept rainwater after maturity, the growing season was extended to delay or prevent the vegetation from reaching maturity. A value of 3,500 PHUs which is the highest allowable value for this parameter in the model was used for the grassland, mixed forest, tea and shrub land cover types.

Table 9-6: Modification of the management files for the perennial land cover classes, (data source Breuer et al., 2003)

Variable	Land cover class		
	Forest	Shrub	Grass
IGRO	1	1	1
Plant _ID	7	6	6
LAI_INIT	4	2.5	3
BIO_INIT (kg/ha)	50000	30000	30000
PHU_LT	3500	3500	3500

After implementing these changes, the biomass and the crop yields for the perennial land cover classes increased considerably, but the water balance were not yet satisfactorily improved. While the statistical performance was good with NSE = 0.61, the modelled water yields, and the ET were underestimated, Table 9 7.

Table 9-7: Model performance with perenial HRUs modification

Variable	Simulated	Observed(expected)
Water yield, mm	171	~234
Evapotranspiration,mm	624	>700
NSE	0.61	>0.50

The next step of the model improvement consisted of the use of a crop level land use map. In the lower sections of the study area, there are two distinct rainy seasons (MAM and OND) resulting in two peaks in the vegetation growth. In the default management file, the growing season is scheduled by heat units (HU). Crop growth only occurs on those days when the mean daily temperature exceeds the base temperature. Crop growth only occurs on those days when the mean daily temperature exceeds the base temperature. The heat unit accumulation for a given day is calculated with the equation;

$$HU = \overline{T_{av}} - T_{base} \ when \ \overline{T_{av}} > T_{base} \ \text{-----------------------------} 9.6$$

Where: $\overline{T_{av}}$ is the mean daily temprature ($^{\circ}$C), T_{base} is the plant's base or minimum temperature for growth ($^{\circ}$C).

The total number of heat units also referred to as potential heat units (PHU) required for a plant to reach maturity is calculated as

$$PHU = \sum_{d=1}^{m} HU \ \text{---------------------------------------} 9.7$$

Where; d=1 is the day of planting and m is the number of days required for a plant to reach maturity.

This means that the growing season for the vegetation is dependent on the total heat units needed by the specific crop to accumulate to reach maturity. The growing season starts if the temperatures are above a minimum value for a specified number of days and ends if temperatures drop below a minimum value for a specified number of days. To overcome this, the scheduling of planting and harvesting was done by the use of planting dates for the agricultural land use types. A crop calendar (Table 9-8) was developed from field surveys. Maize growing is the key agricultural activity and exhibits two distinct patterns, depending on the agroclimatic zone in which it is grown.

In the SWAT crop database (crop.dat), a new crop designated as upland corn (and assigned the SWAT code MAIZ) was introduced. Two maize crops classes namely: lowland maize (CORN) and upland maize (MAIZ) are therefore considered in SWAT simulation. The two classes share similar characteristics in the crop.dat file with the only difference being the timing of planting and harvesting. Also in the lowland maize (CORN) growing areas two growing seasons are simulated, while only one season was simulated in the upland maize (MAIZ) areas. Annex 8 shows the editing of the management files for the different maize classes. Table 9-9 represents the full spectrum of changes made on the management file in SWAT to better mimic the actual situation on the ground.The changes in Table 9-9 include those performed on the perennial land cover types (Table 9-6) as well as changing the management files for crop cover types to more realistically control the planting and harvesting cycles.

Table 9-8: The crop calendar for the main crops in the different zones

ZONE		Jan	Feb	Mar	Apr	May	Jun	July	Aug	Sep	Oct	Nov	Dec
Molo	Maize		LP	P	W								H
	Potatoes												
Bomet	Maize	H	LP	P		W		H		LP	P		W
	Potatoes						LP	P	W		H		
Chebunyo	Maize	H	LP	P		W		H		LP	P		W
	S. Potato						LP	P	W	H			
Olulunga	Maize	H	LP	P		W		H		LP	P		W
	Wheat	LP	P		W			H					
	Potatoes							LP	P	W		H	

Abbreviations				colour code		
LP	Preparation	W	Weeding		Green off	H Harvest
P	Planting	H	Harvesting		Green on	

Table 9-9: Modification of the SWAT management files

Code	Description	SWAT code	Modifications
3	Lowland maize	CORN	Plant Mar 1st, Harvest July 30th
			Plant Oct 1st, Harvest Jan 31th
6	Closed shrub	SHRB	LAI_INIT 3, BIO_INIT 30ton/ha,
7	Rainfed shrub crop	RFSC	LAI_INIT 3, BIO_INIT 30ton/ha,
10	Rainfed shrub crop	RFSC	LAI_INIT 3, BIO_INIT 30ton/ha,
11	Forest mixed	FRST	LAI_INIT 4, BIO_INIT 50 ton/ha,
15	Upland maize	MAIZ	Plant March 1st, Harvest Dec 30th
19	Forest mixed	FRST	LAI_INIT 4, BIO_INIT 50 ton/ha,
20	Rainfed tree crop	RFTT	LAI_INIT 3, BIO_INIT 30ton/ha,

The better reflect and control the parameterisation of the changed model the watershed was re-divided into 6 sub-basins. New user defined lowest points for the each sub-basin (labelled A-F) are shown in figure 9-7. Points A and B marks the end of the upland corn growing areas and were selected from the elevation bands. They also define the upper boundaries of the Mau forest. Points C and D are located downstream of the forest and at the site of the stream gauging stations 1LA03 and 1LB02 respectively. Point E is located immediately downstream of the confluence of the Amala and Nyangores rivers. Point F at the site of station 1LA04 marks the end of the human activities including cultivated agriculture, and start of the Maasai Mara natural reserve.

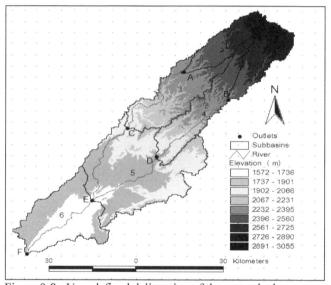

Figure 9-8: User defined delineation of the watershed

The performance of the model simulation for a daily time-step is given in table 9-10 below. The total water yield component of the water balance is within ±5% of the observed fractions. The close approximation of the watershed processes represented by the closeness in the values of the observed and simulated fractions of the water balance indicates a good match. The statistical indices PBIAS = -12.34%, NSE =0.51 and RSR= 0.7 are within the "good", "satisfactory" and "satisfactory" rating respectively according to Moriasi et al. (2007) criterion of NSE > 0.5, PBIAS <±25 and RSR < 0.7 for stream flow. Though the SWAT model is satisfactorily calibrated for flows in this study, the evapotranspiration for the forest and other perennial land cover types was still very low.

Table 9-10: Model performance for annual water balance

	Water balance parameters				Goodness of fit			
	Water yields, mm	Baseflow flow, mm	Surface flow, mm	Baseflow fraction (%)	ET, mm	NSE	PBIAS	RSR
Observed	234	167*	67*	71				
Simulated	210	150	02	61	670	0.51	12	0.7
% diff (O-S/O)	-3	5	-22					

In order to improve on the ET in forest, a modified SWAT version (SWAT_L) was deployed together with other adjustments in the crop database and the parameter maximum canopy storage (CANMX). The plant canopy can significantly affect infiltration, surface runoff and evapotranspiration. The maximum canopy storage (CANMX) also known as interception capacity (Ic) is the maximum amount of water that can be trapped in the canopy. The parameter was changed in the .hru files for all the land cover types. The value used for Ic (Table 9-11) for each cover type was derived from Breuer et al. (2003).

Table 9-11: Typical values for interception capacity (Ic) (Breuer et al., 2003).

Type of biome	crops	Herbs/grass	conf. forest	shrubs
Ic	2.6	1.9	1.9	1.1

To test the modification, the SWAT2005_L was run on the DOS prompt. The ET for tea and shrubs increases drastically. Example ET in Hru1 with tea cover changed from 416mm to 844mm. However the ET for a forest hru (hru4) changed marginally from 489 to 647mm. With the ET in tea hru1 doubling and the ET in forest Hru4 changing only by 32%, the crop database was explored for any mis-representation of realistic values. The most sensitive parameters in the crop data list were analysed. The harvest index for optimal growing conditions (hvsti) and the maximum stomatal conductance (GSI) were found to be the most sensitive. Hvsti defines the fraction of the aboveground biomass that is removed in a harvest operation. The value defines the fraction of plant biomass that is lost from the system and unavailable for conversion to residue. The harvest index for forest has been set at 0.76 and was not changed, since reducing the hvsti would reduce the ET further.

Stomatal conductance of water vapor is used in the Penman-Monteith calculations of maximum plant evapotranspiration. The typical values for the maximum stomatal were found from literature. The maximum stomatal conductance (g) values for tropical forest is 5 mms^{-1} (Schulze et al., 1994). In SWAT crop database the default value is 2 mms-1 and was changed to 7 mms-1. The g value for all the relevant landcover types were changed accordingly. Using the best parameters in the last calibrated model, the model re-run with the new crop database using both the standard SWAT2005 and the modified SWAT 2005_L. The results of the simulations are summarized in Table 9-12. There is a significant reduction in the water yield from both the standard SWAT model with new crop database and the modified SWAT 2005_L, but no significant difference between the interventions (columns b,c,d in Table 9-12). The ET was considerably higher for all the interventions, with the modified SWAT having moderately higher average ET than the standard SWAT.

Table 9-12: Comparison of water balances for the different SWAT versions: a) default crop data and SWAT2005, b) default cropdata with SWAT2005_L, c) new crop data with SWAT2005, d) modified crop data with SWAT2005_L.

	Default	Interventions		
	SWAT2005 mm, (a)	SWAT2005_L Oldcropdata mm, (b)	SWAT2005 Newcropdata mm, (c)	SWAT2005_L Newcropdata mm, (d)
Pcp	1155	1155	1155	1155
Surface runoff	90	80	86	80
lateral soil	31	28	25	24
Shallow AQ	120	68	64	52
Revap	37	37	37	37
Deep AQ	35	24	23	20
Total AQ Recharge	354	235	230	204
Water Yied	241	176	175	156
ET	679	814	817	847
ET_Forest	416	844	827	852
ET_Tea	489	648	825	872

The model was re-calibrated to generate new best parameters sets for the modified model. With the modified SWAT, the simulated ET for the forest and other perennial landcover types is comparable to the other annual cover types with the modified crop database. The modified SWAT however has a poorly balanced hydrology with no surface runoff, column b (Table 9-13).

The modified SWAT version shuts out the LAI loop and thus affects the distribution to the surface runoff routine fraction. The algorithm was however instrumental in identifying the deficit in the default crop database to simulate ET for tropical forests and woodland. When the crop database is changed, the modified SWAT_L has little influential advantage on the ET, indicating that the modification has similar positive effect on the ET, but a negative/undesirable effect on the surface runoff. The ET for the perennial cover types was high, although still lower than the annual landcover types like maize and pasture. To address the shortcoming of zero surface runoff, the standard SWAT with the new crop database was adopted.

Although the SWAT manual (Nietsch et al., 2007) recommends the initialising of the PHU_LT when IGRO is 1, the SWAT model was simulated without making this file modification to allow Forest transpire without limitation. The resulting recalibrated model is given in column c, Table 9-13. The forest ET is now the high, although the surface runoff is low (15% of flow). Manual calibration was used to fine tune the calibration and improve on the water balance fractions especially the surface runoff, column d. The calibrated parameters for the adapted SWAT model are provided in table 9.14.

Table 9-13: Comparison of recalibrated SWAT model versions for forest ET optimisation.

Fraction	SWAT2005	SWAT2005_L	NO PHU mm,	NO PHU with Manual tuning mm, (d)
mm	mm, (a)	mm, (b)	(c)	
Pcp	1155	1155	1155	1155
Surface runoff	80	0	33	80
lateral soil	30	101	77	30
Shallow AQ	100	115	142	123
Revap	37	29	32	35
Deep AQ	24	28	2	2
Total AQ Recharge	234	280	280	255
Water Yied	210	216	252	233
ET	816	771	765	793
ET_Forest	824	739	1032	1038
ET_Tea	825	799	828	838

Table 9-14: Minimum and maximum range of SWAT parameters and the initial and best parameters values.

	min	max	initial Value	Final value
		parameter values		
Sol_Awc	-50	50	0.1 - 0.9	0.06 - 0.57
Cn2	30	100	40 - 83	42.24 - 87.64
Esco	0	1	0	0.54
Sol_K	-50	50	13.19 - 45.61	19.52 - 67.50
Rchrg_Dp	0	0.1	0.05	0.10
Epco	0	1	0	0.71
Ch_K2	0	150	0	0.39
Alpha_Bf	0	1	0	0.00
Surlag	0	10	0	0.09
Canmx	0	10	0	8.19
Sol_Alb	0	0.1	0	0.08
Gw_Delay	0	50	31	0.14
Gwqmn	0	200	0	81.72
Blai	0	1	0	0.79

The resulting calibrated model has simulated water yield very close to the observed flow (±1mm), surface runoff at 35% of the flow. According to the results of the baseflow separation analysis, surface runoff should contribute to between 30 - 40% of the total water yield. The statistical parameters for the calibrated model were -8%, 0.55, and 0.67 indicating "very good", "satisfactory" and "satisfactory" for the PBIAS, NSE and RSR respectively (Table 9-15)

Table 9-15: Model performance for assessed annual water balance fractions.

	Water balance parameters					Goodness of fit		
	Water yields, mm	Baseflow flow, mm	Surface flow, mm	Baseflow fraction (%)	ET, mm	NSE	PBIAS	RSR
Observed	234	167*	67*	71				
Simulated	233	150	83	64	793	0.55	-8	0.67
% diff (O-S/O)	0	6	-24					

*calculated from observed data (baseflow separation)

From the representation of simulated ET in Fig 9-8, the ET for the forest is on average higher than the rest of the land cover types. This is consistent with expected ET relationship between different biomes (Cho et al., 2011). The lowland corn (CORN) has the highest variability in ET due to the wide range of agroclimatic zones in which its is grown. The rainfed shrub crop (RFSC) which in reality represents small scale tea farms interspiced with crops and agroforestry has higher ET than the range land bush (SHRB) due to the higher water availability in the sub-humid areas. The main driving force in the ET is the water availability. Vegetations types growing in the humid and semihumid zones have higher ET than those in semi-arid areas.

The SWAT model was validated with streamflow data fro 1975-1977. The NSE for the validation was 0.52. The lower NSE for the validation stage as compared to the calibration NSE could be attributed to the available data as shown in the graphic representation of the time series (Fig 9-10)

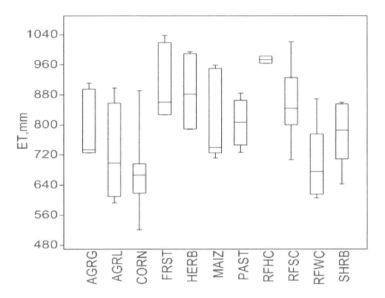

Figure 9-9: The variability in ET for the different land cover classes as simulated by the SWAT model

A satisfactory simulation of the river flows does not necessarily indicate a satisfactory simulation of the individual hydrological processes, due to the equifinality (Beven, 2001) issues for model input parameter values. 'Internal' model performance indicators need to be assessed to demonstrate that the model is also providing a good hydrological simulation (Glavan, 2011). The SWAT_checker stand alone tool (White et al., 2012) was used to assess the hydrological balance of the model simulation. The tool compares simulated outputs and typical value ranges from a database. The model reads selected SWAT output, alerts the user of values outside the typical range, and creates process-based figures for visualization of the appropriateness of output values. The tool's check for this study returned no error warning, indicating that the hydrological components were well within expected typical ranges. Figure 9-9 shows the output of the SWAT_checker analysis of the calibrated model output for the study area.

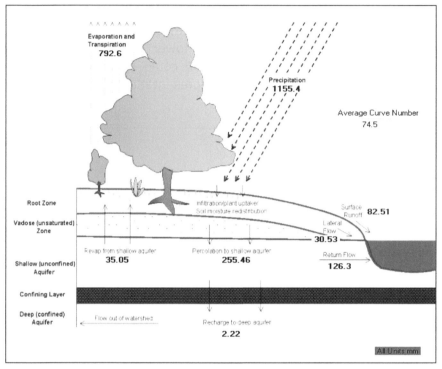

Figure 9-10: Visualisation of the hydrological cycle components for the SWAT modelling of the Upper Mara river basin.

The graphical representation of the observed and the simulated flow is given in Fig 9-10. Both the observed and simulated flows respond well to the rainfall events. After the initial warm-up, the simulated flow corresponds fairly well to the observed flows except for the peak flows. The highest residuals in the time-series plot seems to coincide with the period in the flow observations when there were peak flows in 1971 and 1974 (Fig.9-10), indicating a model failure to match peak flows. Several studies have explained the failure by SWAT to effectively simulate peak flows.

According to Borah et al., (2007) and Mehmet et al., (2009), the SWAT model has an inherent shortcoming in predicting peaks. SWAT has the capability to accurately capture the distribution of daily runoff over a long period. Since SWAT is not a single event model it has difficulties to predict some extreme flow events. Also due to the representation of rainfall with a limited number of stations and the sub daily variations in rainfall intensity, the peaks cannot be captured by the SCS method on the daily time step (Gao and Long, 2008).

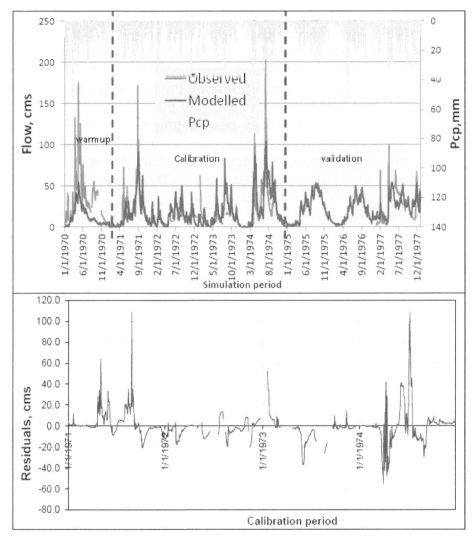

Figure 9-11: Time series of the simulated and the observed (top) and residuals for the calibration period 1971-1974 (bottom)

Since rainfall was the only variable observed within the study area, the large number of missing data both in the input climatic variables (i.e. temperature, wind and solar radiation and relative humidity), and in the observed flow time series presents a challenge in model forcing during autocalibration. The use of the neighbourhood stations and the model weather generator could also be a source of uncertainties.

The model performance improvement procedures outlined above used recalibration of the same range of model parameters. The performance of the different stages in the improvement process was also compared without recalibration using the best parameters of the latest improvement (top down approach). The change in the water balances for the six different improvements was compared with the observed flow data (Table.9-16). All initial simulation using the FAO map was found to relate closer to the final simulation interms of the water yield. This land use classes in the FAO map have been aggregated to large classes similar to the developed map. Besides controlling the parameterisation process, land use maps play an important rule in the the water balance due to the use oft the Curve number method. The drop in the average catchment ET after the introduction of the interpolated rainfall was an indication that areal rainfall may not be best suited for simulation of catchment processes. Though the problem was attributed to the decrease in the precipitation input (1432 to 1155mm), the low precipitation itself was a result of the interpolation process. This implies that the use of satellite gridded data with a coarser resolution (than the interpolated data) is less likely to give a good representation of the basin processes. Observed data, therefore, remain a vital component of hydrological models until finer resolution gridded data is available.

9.7 The SWAT model validation

9.7.1 Single point validation

The SWAT model was validated using the classic hydrological model validation process, based on stream flow data from the period 1975 - 1977. The data available for the validation stage was poor (with 24-100% gaps); with only 1977 has less (24%) missing data points, compared to 70% and 100% for 1975 and 1976 respectively. Against this background of poor observed data, further validation of the model using distributed variables was explored in this study.

166

Table 9-16: Water balance components in different improvement procedures the SWAT modelling.

a). With re-calibration					
SWAT version	Water Yield,mm	Surface Runoff,mm	Baseflow mm	ET	pcp
Observed	**234**	**67**			
1 Initial setup (FAOmap)	287	0	286	976	1432
2 1976-Lulc map	188	9	180	980	1289
3 Interpolated rainfall	197	1	196	708	1155
4 Modify perenials	171	52	126	578	1155
5 NDVI map	171	0	171	624	1155
6 NDVI + All mdgt files	241	92	158	679	1155
7 New cropdata	233	83	157	793	1155
b). Without recalibration - change with best simulation first					
1 Initial setup (FAOmap)	218	46	183	1014	1432
2 1976-Lulc map	305	147	173	852	1289
3 Interpolated rainfall	209	70	149	768	1155
4 Modify perenials	203	73	139	776	1155
5 NDVI map	198	90	119	820	1155
6 NDVI + All mdgt files	277	91	196	678	1155
7 New cropdata	233	83	157	793	1155

9.7.2 Distributed validation

Single outlet calibration and validation of hydrological models have been faulted for not fully accounting for the spatial variability in the modelled basin (Qi and Grundwald, 2005; Rahbeh et al., 2011). They suggested a spatial approach by the calibration and validation of SWAT at several gauge stations of sub-watersheds within a larger watershed. The major shortcoming of rainfall-runoff modelling, particularly in ungauged basins, is the lack of long-term stream flow observations with sufficient spatial coverage that would allow for adequate model calibration and validation (Miller et al., 2002). The International Association of Hydrological Sciences (IAHS) defined an ungauged basin as "one with inadequate records of hydrological observations to enable computation of hydrological variables of interest at the appropriate spatial and temporal scales, and to the accuracy acceptable for practical applications" (Sivapalan et al., 2003). The quantification of the hydrological budget is extremely difficult over large spatial domains and over large time periods, as in situ observations are labour intensive and expensive to generate (Lakshmi, 2004).

According to Sivapalan et al., (2003) and Srinivasan et al., (2010) different methods have been used to build hydrologic modelling systems in ungauged basins, including the extrapolation of response information from gauged to ungauged basins, measurements by remote sensing, the application of process based hydrological models in which climate inputs are specified or measured, and the application of combined meteorological hydrological models that do not require the user to specify precipitation inputs. Other variables that have gained increased use in spatial model calibration include the evapotranspiration (ET), biomass, the leaf area index (LAI) and the crop yield. LAI represents the size of the interface between the plant and the atmosphere for energy and mass exchanges. It is thus of prime interest for the energy balance, photosynthesis, transpiration and litter production. LAI could be used to validate canopy photosynthesis models which simulate the growth and the canopy development based on climate and environmental factors (Baret et al., 2006). In SWAT, the LAI is simulated as a function of heat units using the following series of functions

$$fr_{PHU} = \frac{\sum_{i=1}^{d} HUi}{PHU} \quad \text{------------------------9.8}$$

$$fr_{LAImx} = \frac{fr_{PHU}}{fr_{PHU} + exp\,(li - l2.fr_{PHU}} \quad \text{------------------------9.9}$$

$$\Delta LAI_i = (fr_{LAImxi} - fr_{LAImxi-1}).LAI_{mx} \cdot (1 - exp\,(5.(LAI_{i-1} - LAI_{mx}) \quad \text{-------9.10}$$

$$LAI_i = LAI_{i-1} + \Delta LAI_i \quad \text{------------------------9.11}$$

The LAI during the senescence period is given by:

$$LAI = LAI_{mx} \cdot \frac{(1 - fr_{PHU})}{(1 - fr_{PHUsen})} \quad \text{------------------------9.12}$$

Where; fr_{PHU} is the fraction of potential heat unit at point i , HUi is the heat unit, PHU is the potential heat unit and fr_{LAImx} is heat unit at LAImax, which is maximum Leaf Area index (BLAI)

Crop yield or biomass generally depends on evapotranspiration and on soil moisture, and can therefore be used as an alternative for evaluating the combined actual evapotranspiration (AET) and soil moisture within the hydrological budget (Srinivasan et al., 2010). These indices are either measured in the field or generated from remote sensing. Lakshmi (2004) noted that satellite data represent a wealth of information, which can bridge the gap between point measurements and computer-based simulations, and that larger basins (100–10 000 km^2) are perfect locations for the use of satellite and radar data, as they will have multiple pixel coverage. Satellite remote sensing is an attractive tool for crop area and Net Primary Productivity (NPP) estimates because it provides spatial and temporal information on the location and state of crop canopies (Moulin et al., 1998).

In our case, a distributed validation was necessitated by the fact that the 2905 km^2 watershed has complex hydrographic and agroclimatic profiles, making the use of one monitoring site ineffectual. Furthermore, the river gauging stations available have data of questionable quality (Gann et al., 2006). The lack of spatial reference for the HRU presents a problem in presenting distributed information. While HRU approach can indicate the effect of management practices within the HRU, it fails to show the interaction between the HRUs as they are not internally linked within the landscape but are all routed individually to the basin outlet. Therefore, the impact of management of an upslope HRU on a downslope HRU cannot be assessed. The lumped method of using dominant soil and land use and the HRU delineation do not consider landscape position when computing runoff (Arnold et al., 2010).

In the absence of data for model calibration, quantification of output uncertainty due to spatial representation of the watershed input data should be assessed and minimized to appropriately interpret modeling results (Migliaccio and Chaubey, 2008). In addition, the SWAT was developed to be used in ungauged watersheds. In order to resolve the spatial location shortcoming, a new term referred to as land use soil unit (LUSU) was introduced in this study to represent a physical location on the ground. A specific combination of soil and land use, with a given land use layer overlaying the soil layer. The polygon overlay, intersected the two data layers producing new features, and with combined the attributes of intersecting polygons. Prior to the overlay procedures, the datasets were properly geo-registered. Data layers were referenced to the same coordinate system, the same map projection, the same datum, and the same unit of measure.

The differences between the LUSU and the HRU are; 1. A space in the watershed other than a point has a unique land use and soil type. 2. There are fewer LUSUs than HRUs in a watershed, since HRU is defined in the sub-basin, while LUSU is basin wide (i.e. repeat HRU are not considered in LUSU). The amalgamation of different HRUs produces smooth units unlike the fuzzy outputs from the HRU delineation (Fig 9-11). The purpose of the LUSU is to make it possible to assess the watershed response to hydrological processes at any point by use of measurable spatially distributed metrics. Physical metrics for yields, biomass, evapotranspiration (ET) and LAI are readily available at a given point from satellite imagery and from data collected from the field on a relatively shorter timeframe than stream flow data. The LUSU system produced 50 units instead of 75 HRUs. The land use classes under crops are used in this study for the comparison of the measured and simulated yield and LAI.

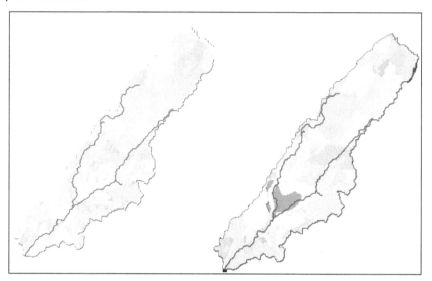

Figure 9-12: Comparison of the Mara watershed decomposition into HRU (left) and LUSU (right)

For the distributed validation of the SWAT modelling results, remote sensed Leaf Area Index (RS-LAI) (§5.5) and the crop yield data gathered from field survey (§5.6) were used. In order to more precisely simulate the actual activities on the ground, two SWAT simulations were used. The default SWAT simulation which represents the calibrated SWAT model (§9.4) and is referred to as unfertilized simulation. Also a SWAT model simulation (fertilized) where the management practices implemented by farmers (from data gathered in the field survey) are incorporated.

9.7.2.1 Validation of the model with remote sensed LAI

The LAI time series for the different agricultural LUSUs corresponding to the period when the land use maps were made (2008-2010) were extracted from the remote sensed VGT4Africa LAI maps (§5.6). They were compared with the LAI values obtained from the SWAT simulation

(Fig 9-12). Whereas the remote sensed LAI (RS-LAI) captures all the green activity on the ground, the SWAT model simulates plant growth for only a single crop at a time. The SWAT model has a definite start-stop sequence which is timed on the calendar year. While the model will reduce the amount of biomass depending on water stress, the plant and harvest periods are pre-determined and do not follow the availability of water. RS_LAI on the other hand reflects the actual vegetation response, as influenced by the moisture content and other conducive environmental factors. The shape of the RS_LAI profile is therefore very random. The shape of the graphs for both remote sensed and simulated LAI has a clear seasonality that correctly represents the phenological profile of agricultural crops.

The SWAT model is able to correctly predict the timing of the start of the growing season. This is critical because it indicates that the model is properly setup to respond to the change in the available water. Depending on the available water, the growing period of the RS_LAI is either the same as that of the SWAT-LAI or longer. The model was able to correctly lag the growing profile for the upland corn crop (MAIZ) in a way similar as predicted by the RS_LAI (Fig 9-12). Although the RS_LAI values are higher than SWAT values, they resonate well with literature values obtained from field studies in the region (Mburu et al., 2011). Remote sensing data may therefore be used for the validation of the start of the growing season in SWAT. The RS_LAI has minimum LAI value which maybe attributed to background LAI. When this is filtered out, the resultant corrected RS_LAI has compared magnitude to the SWAT_LAI in the initial stages. In subsequent seasons the corrected RS_LAI is higher while the SWAT_LAI remains constant, indicating the SWAT models inability to dynamically change with changing environmental (landcover) conditions.

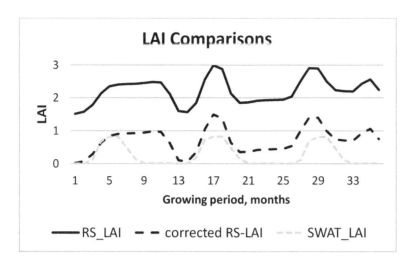

Figure 9-13: Leaf area index (LAI) values for agricultural HRU (CORN) for a scenario with fertilizer and corresponding remotely sensed LAI.

9.7.2.2 Validation of the model with crop yields

The performance on the SWAT model was assessed using both default yields without and also with fertilizer application. From sampled farmers, the commonly used fertilizer was Di-Ammonium Phosphate (DAP) (18:46:0), with application rate ranging from 9 to 247 kg/ha, a mean of 116 kg/ha and standard deviation of 65 kg/ha. Since fertilizer in Kenya is packaged in 50kg bags, 100kg (2 bags) was selected as the nominal rate. For unfertilized simulation, there is no statistical difference (p=0.05) between the different agricultural classes. However there is a significant difference in yields between soil types. The simulated yields are lower than the measured yields except for two soil types (F17 and Up2) whose simulated yields are comparable to observed yields. On an annual average, the unfertilized scenario has water stress (W_STRS) for as many as 50 days and up to 144 days of Nitrogen stress (N_STRS). There is a significant increase in yields with fertilizer application(Fig 9-13). The simulated yields for maize crop under fertilized conditions approximate closely to the observed yields for most of the soil types. In 38% of the LUSUs the application of the 100 kg/Ha was not enough to simulate the observed yields. The yields from two soil types (F17 and Up2) was much higher in the fertilized simulation than the observed data, which is understandable since the initial unfertilised yields were already high. The addition of fertilisers reduced the nitrogen stress by at least half for all soil classes.

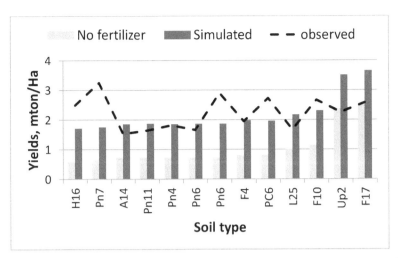

Figure 9-14: Comparison of simulated and the observed lowland corn yields.

The significant difference in yields amongst the soils types in the unfertilized simulation is still maintained after fertilizer application, implying that some soils are naturally more nutrient stressed than others. The response to the fertilizer application was proportionate for all the soil types. A spatial representation of the yields on soil map is shown in Figure 9-14. For all the soil types except Up2 and F17, reasonable yields may only be expected with fertilizer application. The yields increased by 46 to 248% due to application of 100 kg/Ha fertilizer. The increase should not be taken at face value since the model assumes optimum management practices. SWAT-estimated yields represent the typical or potential yield given nutrient, water and temperature conditions. However, in reality, unfavourable external factors like pests, weeds, winds and management practices may exist. These attenuating circumstances may reduce the simulated yields considerably.

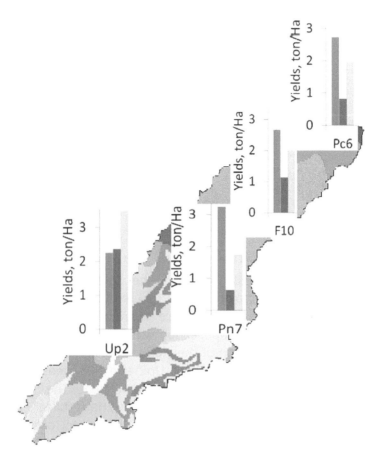

Figure 9-15: Comparison between SWAT simulated yields (red= no fertilizer, green=with fertilizer) and observed field survey (blue) crop yields for selected soil type classes

9.8 Conclusion

The SWAT hydrological model has been successfully adapted for the Upper Mara basin by changing the crop database and the model management file. Input landcover map specially developed for the watershed adaptation was deployed to better simulate farm level activities and enable the distributed validation of the model. The model performance was statistically satisfactory and the water balance fitted well to the observed conditions. The crop yield and leaf area index have been shown as promising methods of SWAT model validation in areas with scarce observed streamflow data.

10. Modeling climatic change impacts on the hydrology of the Upper Mara basin

10.1 Introduction

Tisseuil et al., 2010 described Impact studies of climate change on hydrology as a two-step approach that involves: (i) downscaling GCM outputs to generate local precipitation and temperature, (ii) using the downscaled local precipitation and temperature as input to a hydrological model to project the hydrological changes according to future climate scenarios. The changes in the climatic inputs pathway uses either the downscaling of general circulation models (GCMs) or GCMs coupled with regional climate models (RCMs). Past climate change analysis for Africa indicate that in tropical and subtropical regions more intense and longer droughts have been observed over wider areas since the 1970s, that the frequency of heavy precipitation events has increased over most land areas and that widespread changes in extreme temperatures have been observed over the last 50 years. For future projections in eastern Africa most GCMs are in agreement that temperatures will increase across the region. There is also some agreement that precipitation will increase from December to February in eastern Africa.

Recent trends show a tendency towards greater extremes. Arid or semi-arid areas in northern, western, eastern and parts of southern Africa are becoming steadily drier and there is an increased magnitude and variability of precipitations and storms. Nyong (2005) predicted that by the year 2050, rainfall in sub-Saharan Africa could drop by 10%, leading to major water shortages. De Wit et al. (2006) however suggested that East Africa's future looks better, with increases of drainage density to be expected, because parts of the region may expect an increase in rainfall that could even put it into the wet regime. This is in agreement with Ringius et al., (1996), whose simulations of climate change in Africa indicated that Kenya will be about 1.4°C warmer by the year 2050 (about 0.2°C/decade). Annual rainfall might increase with 20% by 2050 throughout the country, but especially in the highlands, while the potential evapotranspiration (PET) is expected to increase throughout the region by 10% due to the increase in temperature and by about 15% with the inclusion of other climatic changes and changes in the plant physiological characteristics in a CO_2 enriched environment. Dessu and Mellese, (2012) also found that the DJF showed the largest range of change in precipitation for 2050s and 2080s. In "The physical science basis", the Intergovernmental Panel on Climate Change (IPCC, 2007)) working group I states: "Although model results vary, there is a general consensus for wetting in East Africa". The forecasts from the A1B scenario from multiple climate models used for the IPCC (2007) report are shown in fig 10-1. The projection using the Multi Model Datasets scenario A1B for the East African region up to 2099 is summarized in Table 10-1 below

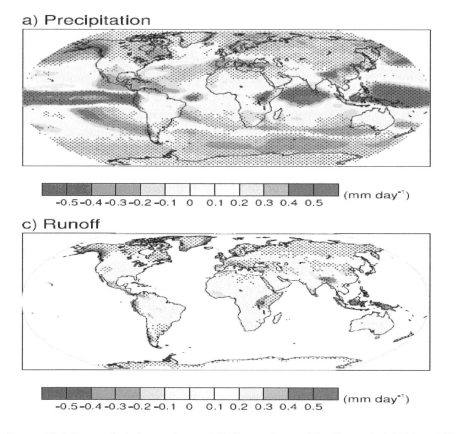

Figure 10-1:Forecasted change in precipitation and runoff for the period 2080 to 2099, as compared to 1980 - 1999 (IPCC, 2007)

Table 10-1: Projected decadal mean change for the A1B scenario through the 21st century for East Africa (IPCC, 2007)

Season	Temperature (°C)			Precipitation (%)			Extreme seasons (%)		
	Min	Max	T yrs	Min	Max	T yrs	Warm	Wet	Dry
DJF	2	4.2	10	-3	33	55	100	25	1
MAM	1.7	4.5	10	-9	20	>100	100	15	4
JJA	1.6	4.7	10	-18	16		100		
SON	1.9	4.3	10	-10	38	95	100	21	3
Annual	1.8	4.3	10	-3	25	60	100	30	1

Legend: T = return period in years

10.2 The climate change sensitivity analysis

In order to assess the impacts of climate change on the hydrology of the Upper Mara river basin, a sensitivity analysis was performed by exposing the SWAT model to variations in temperature, rainfall and CO_2. While exposure relates to the degree of climate stress upon a particular unit analysis, it may be represented as either a long-term change in climate conditions or by a change in climate variability, including the magnitude and frequency of extreme events. IPCC, 2001 defines sensitivity as "the degree to which a system will be affected by, or responsive to climate stimuli, either positively or negatively". Further, according to IPCC 2001, an assessment of the sensitivity of a model to climate change does not necessarily provide a projection of the likely consequences of climate change. Such studies, however, provide valuable insights into the sensitivity of the hydrological systems to changes in climate.

Sensitivity studies of CO_2, temperature and precipitation variations can provide important information regarding the responses and vulnerabilities of different hydrologic systems to climate change, especially in the light of the substantial uncertainty of the GCM climate projections. Possible changes in regional and seasonal patterns of temperature and precipitation and their implications for the hydrologic cycle are as yet poorly understood. An increase of atmospheric CO_2 will directly affect plant transpiration and growth which are inherently tied to the hydrologic cycle. According to Chaplot (2007), an increase in CO_2 - while holding temperature and precipitation constant - will cause increases in water yield. This is due to marked decrease of the stomatal conductance of plants thus decreasing ET, (Wilson et al., 1999). Nash and Gleick (1993) on the other hand reported that higher temperatures lead to increased evaporation rates, reductions in stream flow and increased frequency of droughts. The Precipitation Runoff Modeling System (PRMS) simulated a 30% decrease in runoff with a 10% decrease in precipitation amount in Africa (Legesse et al., 2003). The SWAT simulator has also been used for the simulation of climatic scenarios in hydrological processes. There are two pathways for climate change impacts prediction using SWAT: the land cover/land use change pathway uses either increased atmospheric CO_2 concentrations or plant development and transpiration changes, while the changes in climatic inputs way uses results of downscaled general circulation models (GCMs) or GCMs coupled with regional climate models (RCMs).

In the SWAT simulator, the calculation of the evapotranspiration (ET) takes into account variations of the radiation-use efficiency, the plant growth and plant transpiration due to changes in the atmospheric CO_2 concentrations. The latter is essential for any study of CO_2-induced climate change. SWAT allows adjustment terms, such as the CO_2 concentration, to vary so that the user is able to incorporate GCM projections of atmospheric greenhouse gas concentrations and temperatures into the model simulations.

Although SWAT does not allow incremental increases of atmospheric CO_2 concentration, the impact of the increase of plant productivity and the decrease of plant water requirements due to increasing CO_2 levels are considered (Nietsch et al., 2007). For the estimation of ET, the Penman–Monteith method, which has been modified in SWAT to account for CO_2 impacts, should be used for climate change scenarios that account for changing atmospheric CO_2 levels (Ficklin et al., 2009). The maximum projected global temperature rise by 2099 is 6.4°C (IPCC, 2007). The trend analysis in the study has shown temperature increments of up to 4°C over a 19 year period (1999-2010).

For the sensitivity analysis, the minimum and maximum temperature ranges were set between -5 and 10°C. The maximum range for the sensitivity analysis was selected to accommodate all the possible projected changes including potential uncertanities in the projections.GCM predictions are in disagreement over the direction in which rainfall will go for the East African highlands. Historical trend analyses in the study area show declining rainfall trends. The sensitivity regarding precipitation was conducted with a ±20% range.

Depending on the greenhouse gas emission scenario, atmospheric CO_2 is expected to increase from the current concentration of 400 ppm (Tans and Keeling, 2013) to between approximately 550 and 970 ppm by the end of the 21st century (IPCC, 2001). According to Pritchard et al. (1999), the decrease in stomatal conductance may theoretically be compensated by a potential increase in leaf area with increased atmospheric CO_2. However, SWAT assumes that the leaf area does not increase with increasing CO_2 concentrations. The equation for simulating leaf conductance, with an increased CO_2 concentration in SWAT is given by Easterling et al., (1992) as:

$$g_{CO_2} = g * \left[1.4 - 0.4 * \left({CO_2}/{400} \right) \right] \text{-----------------------------------} 10.1$$

Where: gCO_2 = the conductance modified to reflect CO_2 effects; g = the conductance without the effect of CO_2; CO_2 = the atmospheric CO_2 concentration (ppm); 400 = the present day atmospheric CO_2 concentration

For the sensitivity analysis, historical time series were perturbed, whereby relative changes (percent wise) were effected for the precipitation, while the temperature was increased by absolute values. The climate change scenarios, including the combinations of the different variables, are shown in Table 10-2. The sensitivity was initially assessed by changing individual variables one by one (scenarios 1 to 21) and then by considering different combinations of changes (scenarios 22 to 26). The results from the sensitivity analysis show that an increase in the (minimum or maximum) temperature leads to an increase of the evapotranspiration and a decrease of the water yield (Fig. 10-2). At higher temperature changes, the response of the ET is slightly more sensitive to a change of the maximum temperature than to a change of the minimum temperature. An increase in temperature affects ET primarily by increasing the capacity of the air to hold water vapor. The water yield and the evapotranspiration increase with increased rainfall. Increased precipitation makes more water available for the different fractions of the water balance. Precipitation is assumed to be the only source of water in the basin and the balance will be determined by the input amount. The higher the water input to the SWAT simulator, the higher the expected outputs of water yield and evapotranspiration since the amount that can be stored within the basin is limited.

Table 10-2: Climate change sensitivity scenarios used for the SWAT simulations.

Variable	Scenario	Change
	Baseline	0
	1	-5
minTemp	2	-2.5
°C	3	2.5
	4	5
	5	10
	6	-5
	7	-2.5
maxtemp	8	0
°C	9	2.5
	10	5
	11	10
	12	-20
pcp	13	-10
%	14	-5
	15	5
	16	10
	17	330
CO2	18	490
	19	650
	20	810
	21	970
	22	min-5,max-5,pcp-10
	23	min-2.5,max-2.5,pcp-5
Combined	24	min2.5,max2.5,pcp5
	25	min5,max2.5,pcp-5
	26	min5,max5,pcp10

When the sensitivity of the model to both temperature and the rainfall inputs are simulated together, the water yield and the evapotranspiration increased (Fig 10-3). The increase in ET is higher than with individual climate inputs, due to the accumulation of positive effects. The water yield increased but at a slower rate with the increase in rainfall and temperature.

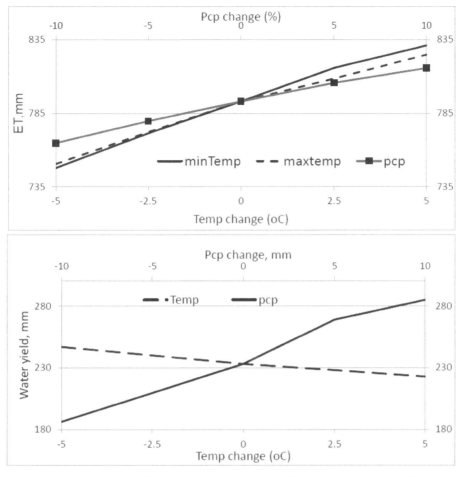

Figure 10-2: Sensitivity of hydrological processes to changing climatic variables applied independently.

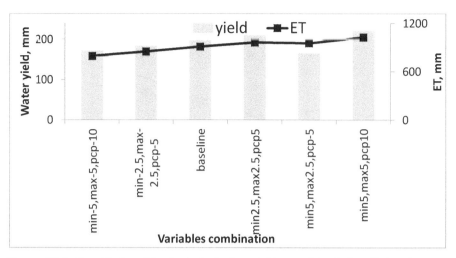

Figure 10-3: Sensitivity of the hydrological model to combined climatic inputs.

The increase in water yield due to increasing rainfall is counterbalanced by the increase of the evapotranspiration due to increasing temperature. At lower temperatures and rainfall, the system is more sensitive to temperature than to rainfall changes. With a rise in rainfall and temperature, the rainfall seems to take over the becomes the bigger driving force in the system. This complicates the assessment of climate change impacts on stream-flow because there is no agreement (only consensus) from the GCMs on the direction of rainfall change for the east African region.

An increase in only CO_2 leads to an increase of the water yield and a decrease of evapotranspiration (Fig.10-4). As can also be seen on the Figure 10-4, the relationship between CO_2 and both ET and the water yield is non-linear, with a sharper response observed at high CO_2 concentrations. These results are consistent with findings by Aber et al. (1995) and Fontaine et al. (2001). There is experimental evidence by Medlyn et et al. (2001) and by Wullschleger et al. (2002) indicating that the stomatal conductance of some plants will decline as the atmospheric CO_2 increases, resulting in a reduction of transpiration.

With less and less transpiration, the total evapotranspiration component of the water balance will decrease with increasing CO_2 build-up. This will lead to a higher water yield in order to maintain the water balance closed. On the other hand, according to Ficklin et al. (2009), the standard SWAT (SWAT2005) does not account for the leaf area increase caused by the rising levels of ambient CO_2. This may lead to an overestimation of ET reduction and of the increase of the water yield.

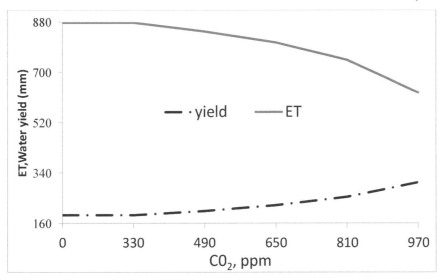

Figure 10-4: Sensitivity of the SWAT model to CO2 variations.

10.3 The weather generator

For making projections of the future situation, daily rainfall amounts, and the daily minimum (Tmin) and maximum (Tmax) temperatures were estimated over a 20-year simulation period using the Long Ashton Research Station weather generator version 5.5 (LARS-WG 5.5) (available from http://www.rothamsted.bbsrc.ac.uk/masmodels/larswg/download.php). LARS-WG can be used for the simulation of daily time-series of climate variables (precipitation, maximum and minimum temperature and solar radiation) at a single site under both current and future climate conditions (Semenov and Barrow, 2002). A weather generator (WG) is a simulator which, after calibration of site parameters with observed weather data at that site, is capable of simulating synthetic time-series of daily weather data that are statistically similar to observed weather (Richardson and Wright, 1984; Wilks and Wilby, 1999). The LARS-WG ver. 5.5 includes climate scenarios based on 15 Global Climate Models (GCMs) which have been used in the IPCC 4th Assessment Report (IPCC, 2007). When compared to other weather generators, the LARS-WG was found to produce better precipitation and minimum and maximum temperature results for diverse climates than the WXGEN and WGN weather generators (Semenov et al., 1998). The LARS-WG also produced better precipitation statistics than the Agriculture and Agri-Food Canada (AAFC) WG (Qian et al., 2004).

The first step in using any stochastic weather generator consists of a statistical analysis of observed weather data for the station in question. According to Semenov et al., (1998), this is important not only for the synthetic data to be similar to the observed data on average, but also for the distribution of the data to be similar across their whole range. The statistical analysis was done by comparing the statistics of the original observed data with those of synthetic data generated by LARS-WG. The χ^2 goodness-of-fit test was used to compare the probability distributions for the lengths of wet and dry series for each season and for the daily distribution of precipitation for each month.

The QTest option in LARS-WG was used to carry out a statistical comparison of the generated weather data with the parameters derived from the observed weather data. In order to ensure that the simulated data probability distributions are close to the true long-term observed distributions for the test site, a large number of years of simulated weather data should be generated (Semenov and Barrow, 2002). Two stations, Narok and Kericho, which have long time (1962-2010) data on rainfall and temperature, were used for the quality test, alongside the station of interest (Centroid station, 1971-1978). The Kolmogorov–Smirnov (K-S) statistic and p-value results of the quality test are listed in Table 10-3.

The LARS_WG generated precipitation for the centroid station (test site) was similar to the observed value (p=1) for both seasonal and monthly time-steps. The Narok precipitation was also well simulated with the generator. The rainfall for June in Narok - and consequently the JJA season – showed the lowest quality with a p value =0.99. The generation of the seasonal precipitation for the Kericho station has a comparatively low statistic in the wet seasons (MAM and JJA) with p-values of 0.41 and 0.43 respectively. Kericho is the wettest of the three stations under review and experiences humid climatic conditions. The statistical test for the generated temperatures was also satisfactory with p-values larger than 0.9 for all tested sites. The results indicate that the WG can be used to generate long term synthetic data series which have the same characteristics as the original time series.

Table 10-3: The quality test statistics for rainfall stations in the study area.

Seasons		Narok K-S*	Narok p-value**	Kericho K-S	Kericho p-value	Centroid K-S	Centroid p-value
DJF	wet	0.06	1.00	0.05	1.00	0.07	1.00
DJF	dry	0.08	1.00	0.10	1.00	0.09	1.00
MAM	wet	0.06	1.00	0.25	0.41	0.11	1.00
MAM	dry	0.07	1.00	0.08	1.00	0.03	1.00
JJA	wet	0.07	1.00	0.25	0.43	0.07	1.00
JJA	dry	0.13	0.99	0.08	1.00	0.05	1.00
SON	wet	0.04	1.00	0.02	1.00	0.07	1.00
SON	dry	0.09	1.00	0.16	0.90	0.06	1.00
Month							
J		0.06	1.00	0.07	1.00	0.04	1.00
F		0.06	1.00	0.05	1.00	0.02	1.00
M		0.08	1.00	0.06	1.00	0.05	1.00
A		0.05	1.00	0.02	1.00	0.05	1.00
M		0.03	1.00	0.06	1.00	0.07	1.00
J		0.13	0.99	0.04	1.00	0.05	1.00
J		0.04	1.00	0.10	1.00	0.01	1.00
A		0.05	1.00	0.06	1.00	0.03	1.00
S		0.05	1.00	0.05	1.00	0.03	1.00
O		0.01	1.00	0.11	1.00	0.03	1.00
N		0.05	1.00	0.08	1.00	0.04	1.00
D		0.05	1.00	0.03	1.00	0.05	1.00

*Kolmogorov - Smirnov

According to IPCC, 1994, a baseline period is needed to define the observed climate with which climate change information is usually combined to create a climate scenario. When using climate model results for scenario construction, the baseline also serves as the reference period from which the modelled future change in climate is calculated. Further the possible criteria for selecting the baseline period include: i). should be representative of the present-day or recent average climate in the study region, and ii), should be of a sufficient duration to encompass a wide range of climatic variations especially anomalies. For this study, we adopted the period 1961-1990 as baseline period. A baseline scenario, without perturbation of the variables of the WG, was generated for the region to represent the current period. This baseline series was used as the reference simulation in the validated SWAT model against which the climate change scenarios are assessed. The baseline data series spans over a period of twenty years, compared to 8 yrs for calibrated model. This long-term period encompasses seasonal, intra-annual, inter-annual and decadal variations.

10.4 The climate change scenarios

IPCC, 2007 decscribes scenarios as "images of the future, or alternative futures. They are neither predictions nor forecasts. Rather each scenario is one alternative image of how the future might unfold. A set of scenarios assists in the understanding of possible future developments of complex systems". In order to explain the scenarios, the IPCC developed several narratives storylines to describe possible future developments. The narrative consider economic development, linkages between societies and environmental consciousness. Four scenarios "families" either at global (A1, B1) or regional (A2, B2) scale were developed. The extent of the historical and predicted global warming as a result of the different scenarios and many GCMs is shown in Fig.10-5.

There are two main possibilities in attempting to predict the future based on models: to choose the best model or to take information from an ensemble of models and combine it into a single estimate. According to Zou and Yang (2004), a single model produces a significant variability due to slight variations in initial conditions or parameter values. The selection of a single climate model run is not viable: initial conditions can never be known to a high enough degree of accuracy and significant uncertainty exists in the values of various physical parameters. Tebaldi et al., (2009) described the structural identity of a model as the way each model discretizes its domain. Haughton (2012) identified the two commonly used procedures to combine GCM results as:

- Unweighted averaging - is the simplest, and involves taking the arithmetic mean of the model outputs. The method is simple and provides a mean as well as estimates of the uncertainty.
- Performance weighting methods - base on the recognition of the fact that no model Is the true model but some models perform better than others due to more accurate algorithms and parameterisations; higher resolution, capturing more complexity; or the inclusion of more physical components. It makes intuitive sense to treat the output of such models with higher regard. Predictions can be adjusted by calculating the performance of each model and weighting better performing models more heavily. Performance (or skill) is generally calculated by some measure of distance between a model run and observations. The difference from observations can be calculated in any number of ways,

depending on the purpose of the experiment. For instance, Rajagopalan et al. (2002,) introduced a Bayesian methodology to determine the optimal weights by using the equi-probable climatological forecast probabilities as a prior.

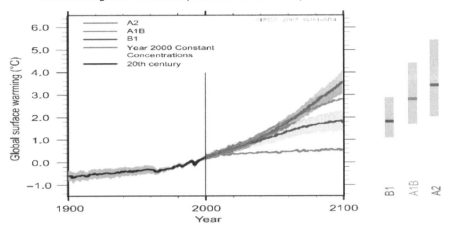

Figure 10-5: Global temperatures in the past and for the future (IPCC, 2007). Solid lines are multi-model global averages of surface warming (relative to 1980–1999) for the scenarios A2, A1B and B1, shown as continuations of the 20th century simulations. Shading denotes the ±1 standard deviation range of individual model annual averages. The orange line is for the experiment where concentrations were held constant at year 2000 values. The grey bars at the right indicate the best estimate (solid line within each bar) and the likely range. The assessment of the best estimate and likely ranges in the grey bars includes the GCMs in the left part of the figure, as well as results from a hierarchy of independent models and observational constraints.

Three IPCC scenarios (A1B, A2 and B1) are available in the LARS-WG database, for the periods 2011-2030, 2046-2065 and 2080-2099. The LARS-WG incorporates predictions from 15 GCMs for the A1B scenario and from 11 GCMs for the B1 scenario. These GCMs - used by the IPCC Fourth Assessment Report (AR4)- form the coordinated set of climate model simulations archived at the Program for Climate Model Diagnosis and Intercomparison (PCMDI) and are know as multi-model data set or MMD of IPCC (2007). Ensembles where developed for the selected scenarios by averaging the results of the different GCMs, the averaging cancels out the individual model's error and decreases the uncertainty.

Annex 9 summarises the centres developing the model, the names of the GCMs, and grid resolution. Annex 10 provides a sample of the climate perturbations with the HADCEM3 GCM. Long term climatic data for the periods 2011-2030 and 2046-2065 were generated and used for the analysis of the climate change impacts on the hydrology of the basin. The period 2011-2030, coincides with national planning strategy, the Kenya "Vision 2030" (GOK, 2007) was particularly used to assess the alignment of projects to the national agenda on climate change.

10.5 The projected rainfall

The mean annual precipitation for all the GCMS in the period 2011-2030 (referred as 2020s) and 2045-2065 (referred as 2050s) are given in Fig. 10-6 and 10-7 respectively. The averaging method was used to establish the ensemble mean for each of the scenarios and time period, since all the GCMs series were generated using the same weather generator. All the ensemble means for the GCMs are higher than the baseline (Fig. 10-8). Most GCMs agree on a wetter climate for the region. Only three of the 15 GCMs have projections lower than the baseline in the A1B 2020 scenario. The A1B scenarios have, on average, higher annual rainfalls than the B1 scenario. The 2050 time period has a higher mean annual precipitation than the 2020s. The variability of the projected precipitation is higher in the A1B_2020 scenario than in the other scenarios.

The deviation of the various GCMs from the baseline for the different scenarios and time horizons can be visualized in Fig. 10-9. Six of the 15 GCMs have a projected rainfall that is higher than the baseline for the two scenarios and time horizons. Only one GCM (HadGEM1) has a projected rainfall that is consistently lower than the baseline for all scenarios. The rest of the GCMs have mixed projections for the different scenarios.

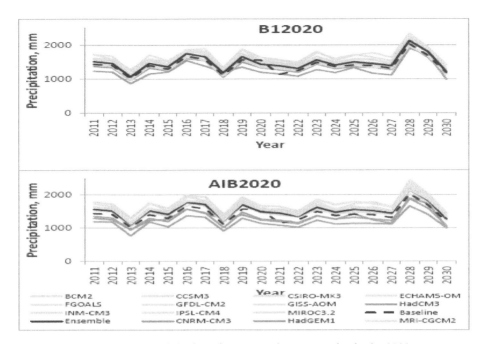

Figure 10-6: Mean annual precipitations for A1B and B1 scenarios in the 2020s

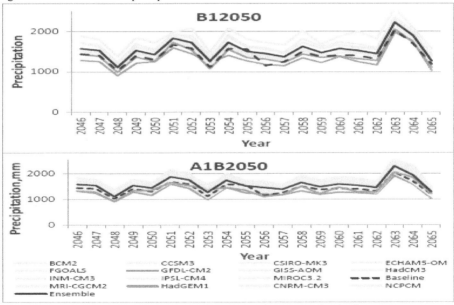

Figure 10-7: Mean annual precipitations for A1B and B1 scenarios in the 2050s

10.6 The projected temperatures

The projected temperatures are higher than the baseline for all the scenarios and time horizons. The highest temperatures rises are projected in the A1B2050 scenarios. All the projections for the 2020s both in scenarios and minimum and maximum temperatures have same magnitude.

The projected changes in both the minimum and maximum temperatures are higher in the 2050s than in the 2030s (Fig. 10-10). The change in minimum temperature is higher than the change in maximum temperatures. On average, the projected change in the 2050s (max.t +1.8 / min.t +2.0^{0}C) is twice as high as for the 2020s (max.t+0.9 / min.t +1.0^{0}C). The highest change in the 2030s is projected in months of August and September while for the 2050s the largest increase in temperatures is expected in June-July.

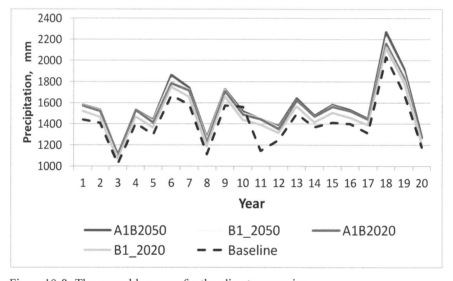

Figure 10-8: The ensemble means for the climate scenarios

190

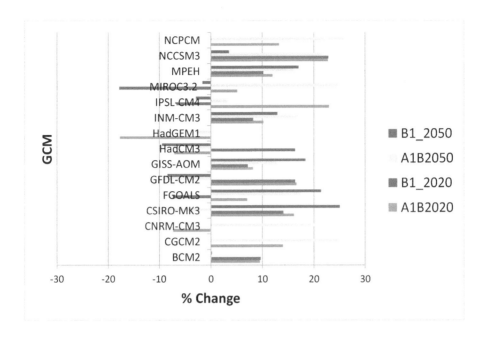

Figure 10-9: Projected percentage of change of the mean annual precipitation with respect to the baseline period (1960-1990)

10.7 The impacts of climate change on the hydrology

The SWAT model was used to assess the impacts of climate changes on the hydrological processes in the Upper Mara basin. Time-series for all the GCMs (15 for A1B and 11 for B1) and the four ensembles were used in the SWAT model to assess the impact on the water yield and on the evapotranspiration. Two options for the perturbations (precipitation only and combined precipitation with minimum and maximum temperature) were explored in order to also test the sensitivity with actual GCMs projected changes.

Fig.10-11 shows that there is no significant statistical difference between the water yields that are simulated based on the 2 perturbation options (Fig. 10-12). The projected water yield (median) is higher than the baseline water yield for all scenarios. The projected water yields are higher in the 2050s than in the 2020s. While the difference in the projected yields between the scenarios A1B and B1 is not statistically significant in the 2020s, the A1B scenario produces, on average, higher water yields than the B1 in the 2050s The variability in the water yield for the different GCMs is highest in the A1B_2050 scenario.

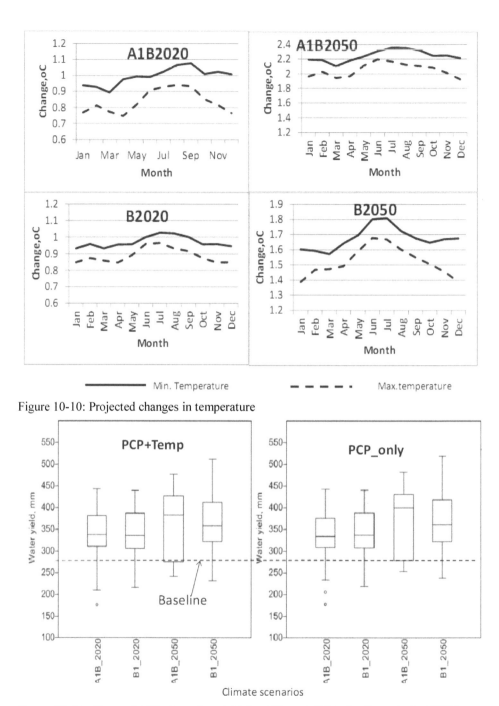

Figure 10-10: Projected changes in temperature

Figure 10-11: The variability in water yield for different GCMs for the selected IPCC scenarios

The combined temperature and precipitation perturbations' option has higher projected ET. Except for the A1B2020s, the rest of the GCMs scenarios have projected higher ET in the combined option than the PCP only. The scenarios in the 2050s have a clear difference in the projected ET from the two options, Fig 10-12. Both the rainfall and temperature are projected to higher in the 2050s compared to the baseline and the 2020s. Since both inputs have a positive effect on the ET, the combination leads to higher difference between the pertubation options. Amongst the Scenarios, the A1B2020 has a markedly high ET compared to the other scenarios, and is the only scenarios with simulated ET higher than the baseline. The ET computation is usually based on the potential evapotranspiration (PET), which is the amount of water that could evaporate and transpire from a vegetated landscape with no restrictions other than the atmospheric demand (Jensen et al., 1990).

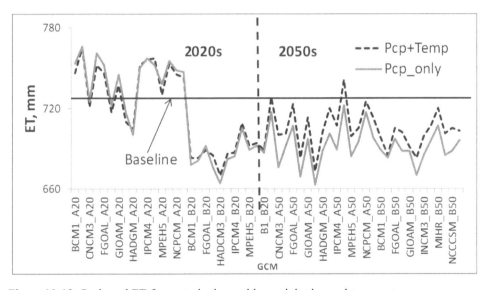

Figure 10-12: Projected ET for perturbations with precipitation and temperature

The rate at which the changing climate affects the water yield was assesed by analyzing the percent change in water yield due to a percent change in precipitation. Regardless of the scenario and the time horizon, the change in the average precipitation between projected and baseline for all the GCMs was plotted against the corresponding change in the water yield (Fig.10-13)

There is a linear relationship between projected precipitation change in the GCMs and the change of the simulated water yield. This linear correlation is exhibited in both perturbations options., the change in water yield from a changing precipitation input may be given by the following the relationship

For the perturbations with precipitation only ($R^2 = 0.96$):

$$\Delta\text{Water yield (\%)} = \lfloor 1.928(\Delta\text{precipitation}(\%) - 7.38) \rfloor \text{------------- 10.2}$$

For the combined perturbations ($R^2 = 0.97$)

$$\Delta\text{Water yield (\%)} = \lfloor 1.924(\Delta\text{precipitation}(\%) - 6.59) \rfloor \text{--------------- 10.3}$$

$$\text{Where; } \Delta = \left[\frac{(\text{Baseline}_{pcp} - \text{projected}_{pcp}) * 100}{\text{Baseline}_{pcp}} \right] \text{---------------------10.4}$$

From Figure 10-13, the highest projected % change in precipitation of 26% for the NCPCM model will lead to a 65% increase in water yield for the Upper Mara. On the other hand the projected 18% decrease in rainfall (for the HadGEM1 and MIROC3.2 models) will lead to a 45% decline in water yield.

No relationship could be established between the projected change in precipitation and the resultant change of ET. The parameterization of the variables affecting the ET is different for the different GCMs under consideration. Whereas the good correlation between the water yield and precipitation makes it possible to approximate the simulated water yield given precipitation, the same direct deductions may not be possible with ET. Whereas for a GCM ranked high in terms of projected rainfall, the expectation is that the water yield will also be comparatively high. For ET, high projected precipitation does not translate into high ET

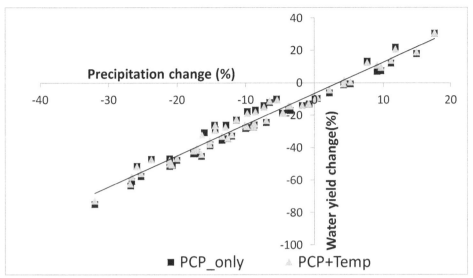

Figure 10-13: Relationship between precipitation input and the resultant water yield for different GCMs

10.8 Bias correction of GCM simulations

According to Ghosh and Mujumdar, the possible reasons for bias in GCM simulations include; partial ignorance about geophysical process, assumptions for numerical modeling, parameterization, and the use empirical Formulae. The two methods available for the correction of the GCM simulation bias are; the change factor (delta change) methods and the statistical bias correction (histogram equilization or quantile mapping) methods. Delta transformation approach was implemented for temperature and precipitation in this study using the transformation developed by Alcamo, 1997 and adapted by Park et al., 2010;

$$T'_{GCM,fut} = T_{GCM} + (T^-_{meas} - T^-_{GCM,his}) \text{-------------------------------}10.5$$

$$P'_{GCM,fut} = P_{GCM} * (P^-_{meas}/ P^-_{GCM,his}) \text{-------------------------------}10.6$$

Where: $T'_{GCM,fut}$ is the corrected temperature, T_{GCM} is the average future ensemble GCM temperature, T^-_{meas} is the measured temperature over the reference period, $T^-_{GCM,his}$ is the average historical GCM temperature, $P'_{GCM,fut}$ is the corrected precipitation, P_{GCM} is the average future ensemble GCM precipitation, P^-_{meas} is the measured precipitation over the reference period, $P^-_{GCM,his}$ is the average historical GCM precipitation.

The delta method ensure that historical data and GCM output have similar statistical properties (Droogcr and Acrts, 2005, Park ct al., 2010, Chen et al., 2011) and transfers the mean monthly change signal between historical GCM and GCM projection period to an observed time series (Eisner et al., 2012). The historical GCM data are model backcasting output and not true historical records. The historical GCM was downloaded using the rWBclimate tool (Arel-Bundock, 2013) from the API climate database which is derived from the same 15 IPCC GCMs used in the LARS-WG.

Besides bias correction various approaches can been used to study and forecast monthly and seasonal rainfall. These methods can be broadly classified into two categories: empirical and dynamical (HE etal., 2013).The empirical methods include statistical models and artificial Neural networks. The forecast lead time of these studies varies from 1 month to 3 months. This short lead time is likely owing to weakened correlation between precipitation and predictor variables with an increasing lag. Persistent exploration of precipitation teleconnections with large-scale climate signals (e.g the El Nino Southern Oscillation, ENSO), result in a large number of predictors as well as a variety of fitting techniques. Forecasts of climate variables, of which rainfall is the most important in water resources applications, can be obtained from Numerical Weather Prediction (NWP) models, which are routinely run by weather services. Although NWP models can provide reasonable large scale forecasts out of one week or longer, because of their coarse spatial resolution they have limited skill in forecasting point or drainage basin-scale rainfall.

10.9 The impacts of climate change on the water resources

Climate change may be perceived most through the manifestation of either low or high extremes. Floods and low-flows are natural phenomena that may hamper the ecosystem functions of the rivers. Low-flows are defined as the flow of water in a stream during prolonged dry weather. They are seasonal phenomena and an integral component of a flow regime of any river (WMO, 1974). High flows information is essential for planning of future water resources and flood protection systems, where system design is traditionally based on the assumption of stationarity of the hydrological variables such as river stage or discharge. The annual maximum flow is hereby often used as a surrogate for floods, recognizing that it does not always represent an out-of-bank flow ((Kundzewicz et al., 2005). There are a number of different ways of analyzing the time series of daily flows to describe low (high) flow regimes. The lowest (highest) mean discharge for 1, 3, 7, 30, and 90 days (minima -maxima) method has been used to characterise low (high) flows in a river system (Riggs, 1972).

The Indicator of Hydrologic Alteration (IHA) methodology uses 32 variables within five groups to quantify hydrological alterations assosciated with presumed perturbations such as dam opearations, flow diversions, or intensive conversion of land uses in a watershed. It works by comparing the hydrologic regimes for pre-impact (baseline conditions) and post impact time frames (Richter et al., 1996). The method allows estimation of the magnitude of impacts but does not enable strong inferences regarding the cause. The magnitude of monthly water condition (monthly median) and the magnitude and duration of annual extremes (minima - maxima) variables are used to assess the alteration due to changes in flow resulting from changes in climate. The Range of Variability (RVA, Ritcher et al., 1997), a modified IHA method introduces target ranges. The RVA target range for each hydrologic parameter is based on selected percentile levels (25 and 75% in this case). The degree to which the RVA target is not attained is a measure of hydrologic alteration and is expressed as a percentage, thus:

$$\text{Hydrologic alteration (HA)} = \left(\frac{\text{Observed frequency} - \text{expected frequency}}{\text{Expected frequency}} \right) \text{--------10.7}$$

$$\text{Range of variability Approach (RVA)} = \frac{(n(\text{in Target)}\,^E/_{nE}) - (n(\text{in Target)}^R/_{nR})}{(n(\text{in Target)}^R/_{nR})} \text{--------10.8}$$

where the observed frequency is the number of years in which the observed value of the hydrologic parameter fell within the target range and the expected frequency is the number of years for which the value is expected to be within the target range. For HA=0, the observed frequency equals the expected frequency; a positive HA indicates that the variable values fell inside the RVA target window more often than expected and vice versa.

The monthly median (IHA group 1), minima and maxima variables (IHA group 2) hydrologic parameters were used to compare the baseline and the projected flows for both the bias corrected and uncorrected A1B2020 and A1B2050 scenarios. The A1B climate scenarios were selected for the impact assessment for both time horizons, since they represent the largest changes in the projected rainfall and temperatures.

The results of the IHA analysis are given in Fig.10-14. All the calculated HA for monthly median flows in the 2020s and 2050s are positive for the uncorrected GCM simulations compared to the baseline. This indicates that median flow the simulated flows have years often falling in the target window for all months of the year, thus no major difference from the baseline period. The bias corrected GCM simulations have the HA negative for the 2020s and the 2050s in comparison to the pre- impact period.   The median flows are out of the 25-75% percentile target window meaning a significant difference in the flow regime due to the impacts of climate change.

Also all the low flows in the 2020s and 2050s were assessed using the minimum flow indices of the 1-day, 3-day, 7-day, 30 day and 90 day minimum. All indices are positive for the uncorrected GCM simulations, and negative for the bias corrected simulations. This indicates that the low flows will often fall within the same target window in the future as they are in the baseline period in the uncorrected situations and outside of the target in the bias corrected. With the ias correction, extreme low flows are expected whcih will be more severe than in the baseline period. As for the high flows, there was mixed signal from the the indices of maximum flow. For the 2020s period all indices except the 1- day and 7-day, are positive for the 2020s. In the 2050s the 1-day, 3- day 7-day are negative compared to the baseline. However the 30-day and 90-day maxima have a positive change. There is therefore higher possibility of experiencing the same level of severity in high flows both in the 2020 and 2050s as currently experienced in the baseline period. However, in the 2050s flood like events are expected in the more 2050s than in the 2020s in reference to the baseline period. The probability of a 1,3, 7 day flood is higher for the 2050s than is in the baseline period.

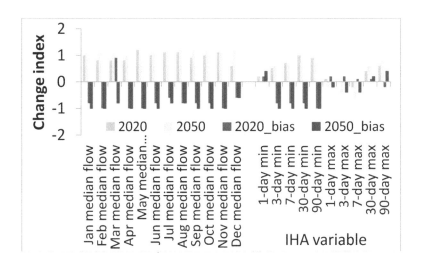

Figure 10-14: Changes in the monthly median, minima and maxima flow indices with reference to the baseline.

In many water engineering applications, the accurate description of extreme surface water states (floodings) and their recurrence rates is of primary importance. The return period, defined as the average number of years to the first occurrence of an event of magnitude greater than a predefined critical event (Benjamin and Cornell, 1970), is an essential design parameter for engineered structures.

The return level is the river flow at a defined return period. The n- year return flood is statistical probability that a certain stream discharge will be equalled or exceeded in n year i.e has n-year recurrence interval. The return period is best predicted if there is sufficient duration of flow records. Holmes and Dinicola, 2010 recommended 10 or more years of data for the determination of recurrence intervals.

The return level at the outlet of the basin for a presumed 30, 50 and 100 yr flood was determined for all the flow scenarios (Table10.4). The return level increases by 11% in the 2020s to 19% in the 2050s. There rise in return flows is consistent with the anticipated hydrological alteration for a I day maximum flood in 2050s. The 1 day maximum for 2020s in the IHA method shows that the flows will fall within the same target window.

When the climate is corrected with historical rainfall the return level flow increases only marginally (by between 1-3 %) for the 30,50 and 100 yr floods. There is therefore minimal change in the return flows from the baseline levels with bias correction.

Table 10-4: Return flow levels for different return periods in the projected and baseline periods.

Scenarios	Return period (yrs)		
	30	50	100
	Return levels (m^3/s)		
Baseline	93	100	110
2020	103	111	123
2020	110	119	131
2020_bias corrected	95	103	113
2050_bias corrected	94	102	113
	% change from baseline		
Baseline - 2020	11	11	12
Baseline - 2050	18	19	19
Baseline- 2020_corrected	2	3	3
Baseline- 2050_corrected	1	2	3

10.9 Possible adaptation strategies

Adaptation is how individuals, groups and natural systems can prepare for and respond to changes in climate, which is crucial in reducing climate change vulnerability (Mitchell and Turner, 2006). Nhemachena and Hassan (2007) classified the farmers' adaptation options as two main kinds of modification in the production systems: a) increased diversification and b) protecting sensitive growth stages by managing the crops to ensure that these critical stages do not coincide with very harsh climatic conditions such as mid-season droughts. Strategies to insure against rainfall variability include: increasing diversification by planting crops that are drought tolerant and/or resistant to temperature stresses; taking full advantage of the available water and making efficient use of it and growing a variety of crops on the same plot or on different plots, thus reducing the risk of complete crop failure since different crops are affected differently by climate events.

Strategies to modify the length of the growing season include: using the additional water from irrigation and water conservation techniques. Adaptive capacity is the potential to adjust in order to minimise negative impacts and maximise any benefits from changes in climate. Integrating climate information into the risk management strategies of communities with climate-sensitive livelihoods depends on effective use of communication infrastructure and networks to support dialogue with users, to facilitate awareness and education campaigns, and to receive feedback so that users can influence the services they receive (Nepad, 2007).

Field-level experiences show that climate adaptation measures are easier to approach from the perspective of programmes rather than through individual activities, thus highlighting the importance of cross-sectoral approaches. Gbetibouo (2009) reported that farmers cited a number of barriers to adaptation, including poverty, lack of access to credit, lack of savings, insecure property rights and lack of markets. Other barriers include lack of information and knowledge of appropriate adaptation measures as barriers to adaptation. Nzuma et al., (2010) summarized the climate change adaptation strategies in "Association for Strengthening Agricultural Research in Eastern and Central Africa" (ASARECA) member countries.

Among the farm-scale strategies adapted for Kenya include: developing and promoting drought-tolerant and early-maturing crop species; exploiting new and renewable energy sources, such as solar power, hydropower and wind power; harvesting rainwater using small check dams; irrigation; reducing the overall livestock numbers by sale or slaughter; delimiting all protected areas to avoid their clearing through encroachment; inaugurating community-based management programmes for forestry, rangelands, national parks; and promoting and strengthening aquaculture, poultry raising, as among other alternative livelihood options (Nzuma et al., 2010).

As a strategy to adapt to the projected high rainfall, this study supports the the construction of a multipurpose dam. The dam will hold the expected excess water during the rainy season and release it for the generation of hydropower, and for use in irrigation during the dry season season. The availability of excess water is demonstrated by use of frequency distribution curves for the baseline and climate scenarios (Fig 10-15). The flows for the ensemble means in A1B2020 and A1B2050 which have the highest rainfall are used for this evaluation. From Figure 10-15, for the uncorrected simulation with the 90% probability of exceedance, the flow has changed from 29.4 m^3/s in the baseline period to 32.4 and 33.8 m^3/s for the 2020s and 2050s respectively. This increase in flow indicates a 10 to 15% rise from the baseline flow. For the bias corrected GCM simulations, there was a 7 and 5% rise in the flow with the 90% probability of exceedance indicating a change in flow to 31.4 and 30.8 m^3/s for the 2020s and 2050s respectively.

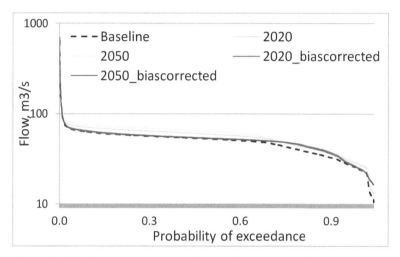

Figure 10-15: Flow duration curves for the flow projected flow due to climate change

Besides the hydrology, changes in temperature and rainfall due to climate change also have socio economic and ecological impacts. A variety of tropical diseases common in Kenya are also sensitive to changes in meteorological parameters such as rainfall, temperature and humidity. These include malaria, cholera, Rift Valley Fever (RVF) and meningitis, among others (ICPAC, 2007). Warmer temperatures will also lead to a decrease of the dissolved oxygen concentrations in the rivers, negatively affecting fisheries and limiting lake overturn (Fick et al., 2005) Climate change is expected to exacerbate the occurrence and intensity of future disease outbreaks and may increase the spread of diseases in some areas (IPCC, 2001).

According to Githeko and Ndegwa (2001), Hay et al. (2002), Zhou et al.(2004), Rainfall and unusually high maximum temperatures are positively correlated with the number of malaria cases in the highland areas of east Africa. Temperature affects the development rates of vectors and parasites while rainfall affects the availability of mosquito breeding sites. There is a correlation between short-term increases in temperature (1971-1995) and malaria incidences, but not on the long term (1911-1995) (Hay et al., 2002). Uncertainties however exist on these correlations. The drop in mosquito control and a rise in drug resistance appear to be confounding studies assessing whether malaria incidence has grown because of - or independently of- climate (Chua, 2012).

10.10 Conclusion

The SWAT hydrological model has been used to analyse the sensitivity of climatic variables for water yield and evapotranspiration. Rainfall was found to be more sensitive than temperature in the Upper Mara basin .

The projected future climate will lead to marginal changes in the streamflow. The incidence of worse low flows is not anticipated. However high freak floods (1,3,7-day flows) are projected especially for the 2050s. Projected higher temperature will lead to higher evaporation rates meaning more loss of water from the basin.

Climate change remains a big threat to livelihood and food security in Africa. When the GCM are no corrected for bias, the projection of the GCMs is for a wetter future in the Upper Mara basin is a little consolation especillay considering that the GCMs resolution is fairly coarse. There is contradiction between the actual historical trends and the projected future precipitation. This contradiction neccesitated the correction of the bias from the GCM simulation with observed and hindcasted GCM simulations. The corrected precipitation and temperatures leads to model outputs which are consistsent with the observed regional tendencies. The climate of the Upper Mara is influenced more by local and regional forcings than the global phenomena. Water resource management strategies should therefore be skewed towards mitigating local causes of hydrological cycle disruptions including calls for more prudent management of the natural resources.

11. Modelling land use and management change impacts on the hydrology of the basin

11.1 The land cover change scenarios

Scenarios of land use changes are generated in order to reflect plausible future land use patterns. Each scenario contains a coherent and internally consistent set of parameters, with reasonable descriptions of possible future states. A scenario is therefore not a prediction of a future state. According to the UNFCCC (2007) "Plausible and credible land use alternatives are developed by taking into account current and historic land use/cover changes; national, local and sectorial policies and regulations; and private activities that influence the use of land in the areas. The level of enforcement of policies and regulations, together with consideration of common practice in the region in which the study is located, are also considered. For identifying realistic land use scenarios, land use records, field surveys, data and feedback from stakeholders, and information from other appropriate sources, including participatory rural appraisal, may be used as appropriate".

The three main pathways commonly used in building different land use scenarios include: the no change scenario, the use of past trends and the use of future trends (Bernoux et al., 2010). In the no change scenario there is no change in the land use or the practices, with respect to the current situation. It represents the most simplistic scenario to build since no additional information is required. However, it does not always reflect the future reality. The use of past trends to get information on the future situation supposes that the changes in land use and practices will evolve in the same way as they have done in the past. The scenario is therefore forecasted using past trends, either using long term (>50 years) or short term (10-30 years) past trends. Recent past trends are especially used to build the baseline for two main reasons. First, in some countries and for some kind of data, the implementation of a monitoring system is quite recent and there is therefore no long term data available. Secondly, the changes of the past 10-30 years are often representative of the current evolution. The use of future trends estimates the future land uses and practices from models based on country planning data. This type of scenario requires the highest number of assumptions on how the situation may eventually evolve. If the models used to build such scenario are robust and fairly reliable, they might logically reflect the future reality. There are many land use models available, which range from very simple to complex. Lambin et al. (2000) reviewed and classified the models based on their methodology and the questions that need to be answered. Such questions include: when and where the changes may occur in the future and what drives the changes or why they change.

For this study, a sustainability first (ecological/environmental concern) scenario was adopted for the land use change impact analysis. In this scenario, regulations regarding the land use are strictly enforced. Statutes, strategic plans and policy documents on the sustainable use of the natural resources are adhered to. These include the decreased forest degradation and afforestation/reforestation to pre-1976 forest cover levels. Observance of riparian laws and banning of farming activities in the riparian reserve is enforced. In the SWAT model build-up, the land use map is replaced with the 1976 land cover map which has only four land cover classes. All the agricultural land cover types are represented as one generic agriculture land cover class (AGRL) as presented in Table 11-1.

Compared to the 2006 baseline landcover, the area under pasture and forest cover changes from 6 and 16% to 18 and 33% respectively. At the same, time the area covered by agriculture decreases from 55% to 28% in this scenario. The least change in cover type is in the shrub class, which decrease marginally from 23% to 21%. Compared to the 2006 baseline map, the water balance fractions changes to varying degrees when the 1976 land map is used (Table 11-4). The average annual water yield reduced marginally (8%) from 233 mm/yr in the baseline to 213 mm/yr, while the evapotranspiration increases by 3% from 793 mm to 813 mm/yr. The.largest relative change concerns a decrease of the surface runoff by 29%. In this scenario, in which the forest cover is increased and land under agriculture is reduced, there is indeed more infiltration due to the slower movement of water over forest, shrub and grassland areas. Under the 1976 land cover scenario, more than 50% of the area is under permanent cover throughout the year. This reduces surface runoff especially at the onset of rains, as agricultural land would otherwise be bare after land preparation.

Table 11-1: Substitution of the baseline 2006 land use map classes with the 1976 land cover map classes.

Name	SWATCODE	main group	% cover	Name	SWAT CODE	% cover
		2006 baseline map			1976 map	
Wheat fields	RFWC	Agriculture	7	Agriculture	AGRL	28
Small cereal	CORN	Agriculture	21			
Small cereals	AGRL	Agriculture	5			
Small herbacious/tea	AGRG	Agriculture	9			
Medium herbacious/trees	RFHC	Agriculture	3			
Small cereals/open trees	MAIZ	Agriculture	4			
Closed herbacious/trees	HERB	Agriculture	6			
		Subtotal	55			
Closed trees	FRST	Forest	12	Forest	FRST	33
tree plantation	FRST	Forest	4			
		Subtotal	16			
Low open shrubs	PAST	Pasture	6	Grass	PAST	18
		Subtotal				
Trees and shrub	SHRB	Shrub	6	Shrub	RNGB	21
Small fields tea	RFSC	Shrub	14			
Large tea fields	RFTT	Shrub	3			
		Subtotal	23			

According to Bruijnzeel 1992, deforestation has a number of potential negative effects on the hydrological processes including: decreased canopy interception of rainfall, decreased transpiration from the replacement vegetation, increased evaporation from the exposed soil surface, decreased soil infiltration because of changes in soil structure, and increased velocity of runoff after removal of surface litter and roughness. Re-afforestation is therefore intented to reverse these negative effects.

With increased infiltration, there is more water available in the soil for plants to take up and transpire. Because the leaf surface area of a forest is generally much higher other cover types, the potential amount of water that a forest can evaporate and transpire is typically much greater than that for other ecosystems under the same moisture conditions. Increasing the vegetation canopy cover affects the water balance through an increase in evaporation, thereby reducing the amount of water available for runoff and stream flow.

The inability of scarce and low vegetation to control the runoff and precipitation turns the hydrological cycle into a sequence of droughts and floods accompanied by extreme winds. (Makarieva and Gorshkov, 2010). The anticipated consequence of more forest cover is the reduction of the extreme flow conditions. Flash floods which occur as a result of high surface runoff during storm events will be reduced. Some of the infiltrated water will still find its way into the river system but at a slower rate.

The results obtained in this scenario are constistent with the results obtained by Githui et. (2009) for the 12000km^2 Nzoia basin also in the Lake Victoria region. The findings on the Upper Mara and those of Githui et al., 2009 are in line with the logical expectations, whereby the removal of vegetative cover generally leads to increased surface runoff. According to Costa et al., 2003, the higher surface albedo, the lower surface aerodynamic roughness, the lower leaf area and the shallower rooting depth of pasture and agriculture compared with forest all contribute to reduced evapotranspiration (ET) and increased long-term discharge. Mango et al., 2012 working on one of the three subbasins (Nyangores subbasin) had a contrary conclusion that who a reduction in forest cover led decrease in water yield.

According to the report of the Government of Kenya taskforce on the rehabilitation of the Mau ecosystem (GOK, 2009) "Continued destruction of the forests is leading to a water crisis: perennial rivers are becoming seasonal, storm flow and downstream flooding are increasing, in some places the aquifer has dropped by 100 meters while wells and springs are drying up". The results for this scenario show that the re-forestation effort will help to ease the pressure on the groundwater water resources. This may be achieved by the increase in the groundwater component of the water balance due to increased infiltration and percolation, as a result longer residence time. The reduction of the surface runoff will lead to decrease in the negative impacts of both soil erosion and floods.

11.2 The management scenarios

Management scenarios relate to the activities undertaken by the occupants of the land, individually or collectively, as a result of a shift in policy or market forces, in order to increase the productivity and profitability. The following management scenarios were separately implemented in the SWAT model and the impacts on the hydrological process and the yields were simulated:

- The application of commercial fertilizers
- The implementation of auto irrigation schedules during periods with water stress
- The change of the cultivated staple crop

11.2.1 The fertilizer application scenario

The fertilizer operation in the SWAT simulator applies both organic and inorganic fertilizer to the soil. The information required in the fertilizer operation include: the timing of application and the type and the amount of fertilizer. This information was obtained from the field survey ($6.3) and corroborated by literature reviews. Field studies also revealed that organic manure is used in the study area, although the use is limited, random and the amounts are unverifiable. The scenario for fertiliser application used in this study involves the application of Di-Ammonium Phosphate (DAP). The month and day for fertilizer application was estimated from the crop calendar (§8.2). A baseline (no change scenario) nominal application rate of 100 kg/ha of DAP (18:46:00) (Nietsch et al., 2007) was used. The 100 kg/ha was set, based on the field survey for a data set of 100 farmers. The surveyed application rate on maize ranges for 10 kg/ha to 248 kg/ha, with a mean of 116 kg/ha and a standard deviation of 66 kg/ha. Also the world average fertilizer application use on maize is 136 kg/ha (FAO, 2006). 18-46-00 represents the percentages N, P_2O_5 and K_2O respectively. The fractions of minerals N and P in the fertilizer are given by the following expressions;

$$N = \%N/100 \qquad \text{---------------------------------------} 11.1$$

$$P = 0.44(\%P_2O_5/100) \qquad \text{------------------------------------} 11.2$$

In this management scenario, the SWAT model is used to assess the impacts of fertiliser use on the maize yields for the different soil types. The sensitivity of the different soil types to fertiliser application was investigated by varying the amount of fertiliser applied. The 100 kg/Ha application rate was used as the baseline.

The study differentiates between three terms commonly used for crop yield as defined by Witt et al., 2009:

- the yield potential - being the theoretical maximum yield of a crop in unlimiting environment both for water and nutrients . Also yield-reducing factors such as pests and diseases are absent and the yield is solely determined by the climate and the germplasm.
- The attainable yield - as the yield achieved with the best management practices, including optimal pest and nutrient management. Although nutrients and pests may not be limiting, soil constraints or water availability may limit the yield.
- The actual yield - often lower than the attainable yield due to constraints like poor crop and nutrient management practices that may also enhance pest and disease pressure.

Due to inputs and uncertainty, it makes little economic sense to strife to close the gap between the potential and attainable yield. However, the gap between actual and attainable yield maybe realistically exploited and significantly narrowed by use of new technologies and implementation of best management practices (Witt et al., 2009)

The Agricultural Sector Development Strategy identifies several interventions to achieve the objectives of the Kenya's Vision 2030, (GoK, 2008). In the Medium Term Investment Plan (MTIP, 2010-2015), maize production is planned to increase form 1.6 ton/ha to 2.2 ton/ha. The goal of this scenario therefore, was assess the quantity fertiliser required to produce a target attainable yield of 3 tons/ha (historical estimates of actual average world yield maize). Maize is targeted for analysis for several reasons including: (1) there is more data available on fertilizer use for maize than any other cereal crops, mostly because of the importance of maize as a staple food crop; (2) maize accounts for nearly 40 percent of all fertilizers applied to cereal crops (IFDC, 2012).

In order to encourage higher productivity through the use of fertilizer, the Kenyan government through the National Cereals and Produce Board (NCPB) has been subsidizing fertilizers to farmers at 70% of the market price. According to Druilhe and Barreiro-Hurlé (2012), "in low input/low output agricultural systems, fertilizer subsidies can play a role in raising fertilizer use and agricultural productivity. They can help demonstrate the benefits of fertilizers and/or kick-start market development by raising input demand at a large scale".

From the sensitivity analysis, the yield response of the all different soils is directly proportional (R^2 = 0.99) to the DAP fertilizer application below 300 kg/ha. Though 300 kg/ha was the highest recorded application rate in the study area (Kilonzo and Obando, 2012), the sensitivity analysis was extended to 500 kg/ha to test the effect of the law of diminishing return on yields. Three distinct soil clusters are indentified, Fig 11-1. The first group consisting of F17 and Up2 soil types and accounting for 12.5% of the Upper mara basin area have a high response to fertiliser application. In this group less than half (50 kg/Ha) of the baseline rate is required to attain the 3 ton/Ha yield target. The relationship between the fertiliser and the yields for these soil types (upto 300kg/ha) is given as:

$$\text{Yields (ton/Ha)} = 0.0108(\text{DAP Fertilizer (Kg/Ha)}) + 2.55 \text{---------------} 11.3$$

The second group in which 77% of the soil types belong, and accounting for 86.5% of the total land mass have a moderate response to fertilizer application. In this group between 150 and 220 kg/Ha of fertiliser is required for the production of 3 ton/Ha of maize. The relationship between the yields and the fertilizer is given by:

$$\text{Yields (ton/Ha)} = 0.0117(\text{DAP Fertilizer (Kg/Ha)}) + 0.81 \text{---------------} 11.4$$

In the last least yielding soil type (H13), 280 kg/ha is required to attain the target yield. This amount is higher than the highest reported application rate. The relationship between the yields and the fertilizer is given by

$$\text{Yields (ton/Ha)} = 0.0092(\text{DAP Fertilizer (Kg/Ha)}) + 0.45 \text{---------------} 11.5$$

The application of more than fertilizer 100kg/ha on the F17 and Up2 soils adds little value to the farmer and not only leads to unnecessarily higher input costs but also to environmental degradation. For most of the soil types, diminishing returns seems to set in at 300kg/ha. The results of the fertilizer analysis also indicate that that the selection of 100kg/ha for the SWAT simulation would lead to under fertilisation of more than 87% of the area in the basin. According to IFA (1992), the fertility demands for grain maize are relatively high and amount, for high-producing varieties, up to about 200 kg/ha N, 50 to 80 kg/ha P and 60 to 100 kg/ha K. The application of 100kg/ha of 18:46:00 DAP fertilizer means only 18kg N and 20kg are applied per hectare. This means that only 9 and 40% of the N and P demands respectively are met. For the targeted attainable yields, the majority of the soil types require 200kg/ha of DAP. At this application rate 82% and 20% of the N and P fertility demands for maize is still unmet. For some soil types like H13, the target yield can not be reached with realistic amounts of fertiliser. For such soil types it is better to use the land for other economic activities like pasture.

Considering all this, there is little chance of impairment of the water quality due to fertilizer use in agricultural land, since the use of the average rates of application does not match the fertility need for 89% of the area. Note that the fertilizer application scenario has not considered the proximity of the fields to the water course or the observance of the buffer zone regulations.

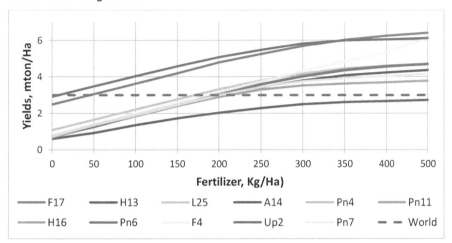

Figure 11-1: Sensitivity of yields to fertilzer application

11.2.2 The auto-irrigation

Irrigation in the SWAT simulator is either scheduled by the user or automatically applied by the simulator in response to a water deficit in the soil. For the first option, the timing and the amount of water that is applied must be specified. For both options, the source of the irrigation water must be specified. Water applied to an HRU is obtained from one of five types of water sources: a reach, a reservoir, a shallow aquifer, a deep aquifer or a source outside the watershed (Nietsch et al., 2007). Currently no reservoirs exist in the study area; however two medium sized multi-purpose dams are planned on the Mara River: at Norera on the Amala tributary in Kenya and at Borenga in Tanzania (NBI, 2011). According to the NBI (2011), the Norera medium dam will provide water for irrigation, domestic water supply, fisheries and flood control.

A reservoir dam was introduced in the SWAT model to represent the Norera dam (Fig. 11-2) and used in this study to supply irrigation water to all the maize HRUs in the subbasins downstream of the reservoir. The Aster DEM was used to delineate the profile of the reservoir and to calculate the surface area covered by the reservoir.

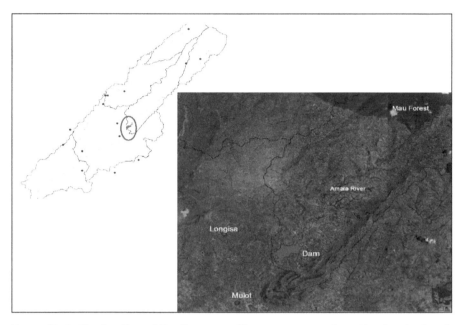

Figure 11-2: The location of the Norera multipurpose reservoir on the Amala river in the Mara River basin

With no more details on the planned dam available from the authorities, reservoir storage characteristics were determined using the Rippl diagram method (Klemes, 1987) and the long-term stream flow data (October 1955-April 1995) from the 1LB02 river gauging station at Kapkimolwa. Information on the monthly water demand for the entire Mara basin was derived from literature values (Gereta et al., 2003; Hoffman, 2007). Hoffmann (2007) estimated the whole Mara river basin annual consumptive demand to be 23 million m^3. The procedure used to determine the reservoir storage volume is described in Annex 11. The highest monthly demand is 2713727 m^3 in July. This demand was used as the worse case scenario for tabulation of the constant demand for a year. The long-term monthly flow was used in the mass curve. The estimated reservoir storage capacity is 432Mm$^{3.}$ and 39Mm^3 for the worse-case and for normal flow conditions, respectively.

In the irrigation scenario only CORN HRUs in subbasins 5 and 6, located downstream of the reservoir, were auto-irrigated during water stress days. Table 11-2 shows the comparison of the irrigated and non irrigated scenarios. Under normal rainfed conditions the CORN land cover types experience water stress of between 7 to 44 days during growing period. The water yield and the ET increased with the introduction of irrigation in the basin. The crop yields increase on average by ca. 100% for the HRUs in the two irrigated subbasins 5 and 6 located downstream of the reservoir.

Table 11-2: Comparison of water balance fractions under irrigated and rainfed conditions.

	Scenario setting	
Variable	Without irrigation	With irrigation
Water yield,mm	233	280
ET,mm	793	800
Revap,mm	35	35
Average crop yield, ton/ha	0.75	1.66
Water stress in HRUs, days	7 to 44	0 -10

11.2.3 The change of cultivated crop type

The main cereal grains consumed in Kenya are maize, wheat, rice, and sorghum. While maize is the most important staple in Kenya, the consumption of wheat and rice has gained prominence in recent years, particularly in urban areas. The consumption of sorghum has traditionally been centered in the drought-prone agricultural areas of Kenya where it is also predominately produced (Nzuma and Sarker, 2010). Maize is the staple food in the study area and the main source of livelihood for the subsistence farmers. Farmers in the study area and the entire Kenyan maize-growing belt suffered heavy losses due to the maize leaf necrosis (MLN) disease in the 2011-2012 growing season. MLN is a serious threat to farmers in the affected areas, who experienced extensive to complete crop loss (Wangai et al., 2012; Adams et al., 2012). According to Wangai et al. (2012), a high incidence of new maize disease was reported in 2011 at lower elevations (1900 masl) in the Longisa division of Bomet County, which later spread to the Narok South district, in Southern Rift Valley, Kenya.

According CIMYT, (2012), the MLN disease is difficult to control for two reasons: 1. it is caused by a combination of two viruses that are difficult to differentiate individually based on visual symptoms, and 2. the insects that transmit the disease-causing viruses may be carried by wind over long distances. Since MLN does not occur on other crops except maize, farmers were advised to avoid growing maize after maize and to grow alternative crops for 2 years in order to reduce losses (CIMYT, 2012). Also, as an adaptation strategy to climate change, sorghum has been promoted as one of the orphaned African crops that need to be grown as an alternative to maize (Schipmann, 2011). In a study to assess the impacts of climate change on crop yields, Ringler et al. (2010) projected a 6% drop in maize and a 4% increase in sorghum yields by 2050. Besides its water saving and nutritional attributes, the economic viability of sorghum growing is higher than that of maize. With an acre under sorghum producing between 20-25 bags (1800-2250 Kg) of sorghum, yields are comparable to those of maize. In Kenya, sorghum retails for approximately 1.5 times more than maize (MOA, 2011). The entry of beer malting companies in contracting farming will push the prices higher and benefit the farmers more.

Against this background, two scenarios where grain sorghum was grown in the watershed were simulated. To avoid the problem occasioned by monoculture cultivation, other agricultural land use types like closed herbacious crops (HERB) and medium to closed herbacious crops (RFHC) will not be changed in the scenario settings. The following scenario's were considered:

1. A scenario where all the lowland maize (CORN), which was severally affected by the MLN, is replaced with the grain sorghum. (Replace_CORN)

2. A scenario where all cereal (agricultural) land use types (notated in SWAT as CORN, MAIZ, AGRG, AGRL) which are at a high risk of MLN attack are replaced by sorghum HRU (Replace_ALL)

Table 11-3 shows the change in the land cover under the different scenario's. The impact of the change in crop types was assessed in terms of simulated changes in the water yield, crop yields and water stress. The results of the simulations show that there is a slight (4%) increase in the water yield for the "Replace_CORN" scenario (Table 11-4). The area under CORN is indeed relatively small (21%) in order to change the water yield for the entire basin. The CORN crop is also limited to the lowland subbasins 5 and 6, located in the semi-arids zones with low rainfall and comparatively low soil fertility.

The "Replace_ALL" scenario, however, has significant impacts on the water yield. The cluster area under this cover types accounts for 39% of the basin area(Table 10-4). For this scenario, the simulated water yield increases by 21% to 282mm/yr from 233mm/yr in the baseline scenario With the introduction of Sorghum, the ET decreases from a high of 793 mm/yr to 774 and 727 mm/yr for the replace_corn and replace_all scenarios respectively. This represents an upto 9% drop in ET compared to the baseline.

Table 11-3: Synthesis of different cover changes under sorghum subsitution scenarios.

Baseline	% cover	Scenarios Replace_CORN	% cover	Replace_ALL	% cover
RFWC	3	RFWC	3	RFWC	3
RFWC	4	RFWC	4	RFWC	4
CORN	21	GRSG	21	GRSG	21
AGRL	5	AGRL	5	GRSG	5
AGRG	9	AGRG	9	GRSG	9
MAIZ	4	MAIZ	4	GRSG	4
HERB	6	HERB	6	HERB	6
RFHC	3	RFHC	3	RFHC	3
Cover under Sorghum (%)	0		21		39

Figure 11-3 compares the flow frequency curves for the replacement scenario's with the calibrated model (baseline) flow data. The curves for the baseline and the "replace with corn" scenario are almost identical except during low flows (>90% exceedance). The "Replace_all" scenarios have flow which are consistently higher than the baseline flow for all the simulation period. Although the observed flows have near zero values, in reality the river does not stop flowing but the extreme low flows are difficult to capture using the staff gauging method.

Sorghum, as a drought resistant crop, has a lower water requirement than maize (Assefa et al., 2010; Zsembeli et al., 2011) and other crops like potatoes and vegetables grown in the study area. The prolific root system, the ability to maintain a stomatal opening at low levels of leaf water potential and the high osmotic adjustment help sorghum to cope with drought. The lower green water fraction (crop water consumption) leads to a higher blue water component of the water balance. The excess water therefore finds its way into the river system through lateral flow. By growing grain sorghum instead of maize, more water will be available for release into the river system. In case of low rainfall, the sorghum crop will be able to withstand drought better than maize leading to food security and safety. With similar rainfall input the number of days with water stress reduced significantly under sorghum cultivation as compared to corn.

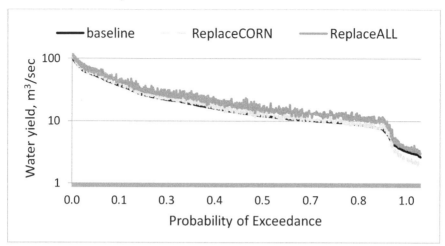

Figure 11-3: The flow frequency curves after replacement of existing crop(s) with sorghum

From Fig 11-4, over a period of 4 years, the average number of water stress days for the most affected HRU (no. 62) reduced from 45 to 28 days, while that of the least affected HRU (no. 30) reduced by half (Fig 11-4B). On average, for all corn HRUs over the same 4 year period, the number of water stress days was reduced from 25 to 10 days (Fig. 11-4C).

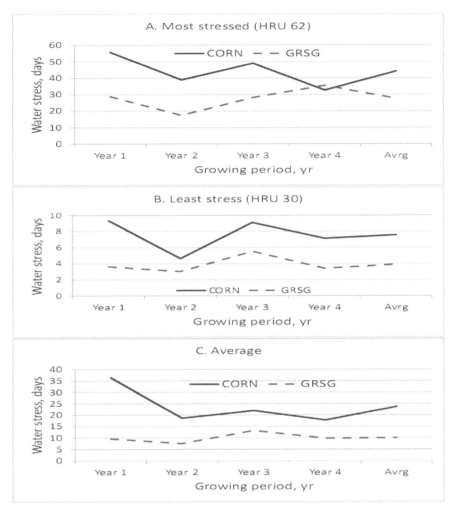

Figure 11-4: Number of water stress days for HRUs under sorghum (GRSG) and maize (CORN) cultivation.

Although sorghum offers the best alternative to maize to ensure grain food security in the larger sub-Saharan Africa region, some factors mitigate against embracing it as an adaptation strategy for climate change. The study area is not classified as a sorghum producing zone as per the Sorghum Atlas fig 11-5 (Wortmann, et al., 2009). Due to the fact that Kenya is the least producer of grain sorghum compared to other countries in the region, and in spite of the water resources related benefits, a paradigm shift in the socio-economic aspects of sorghum adoption needs to be addressed. The conversion of the local farmers to not only produce but also to consume more sorghum products is an uphill anthropological challenge which should be factored in the scenario intervention. In addition to the acceptability challenge, the quelea birds are a big stumbling block to the realisation of good sorghum yields. According to Markula et al. (2009), the queleas also referred to as 'feathered locusts' form nomadic super-colonies of up to 30 million birds, feeding on ripe sorghum, wheat, barley, rice, sunflowers and corn. A flock of 2 million can consume 50 tonnes of grain in a day. Queleas are one of the most abundant and destructive birds in the world, causing $US70 million damage to grain crops per annum (Markula et al., 2009). Traditional methods of control is mainly through bird-scaring are cost effective, and more environmentally friendly, but hugely labour intensive and time-consuming.

11.3 Conclusion

The considered scenarios show that land use changes may affect the water yield and the water resources in different ways. The reintroduction of forest cover improves the water retention in the watershed, increases the ground water recharge, and leads to less surface runoff. Although there is less water yield to the streams, the naturally regulated water flow through increased the proportion of water in soil which moves in form lateral flow, and will potentially lead to fewer occurrences of extreme events in both low and high flows. Despite the existence of a policy framework in favour of this scenario, the heightened political climate surrounding the Mau forest and increased demographic pressure mitigates against the implementation of the scenario.

Figure 11-5.Atlas of Sorghum Production in Eastern and Southern Africa (Wortmann et al., 2008)

The substitution of maize crop with grain sorghum could lead to more water yield in the streams as the sorghum crop uses less water for the same yields. With impending climate change and trends in the study area indicating lower rainfall and higher temperatures, there is a need to shift to more drought tolerant crops. The biggest challenges for this scenario is the behavioural change in eating habits and destruction from quelea birds. The transition from using maize as the staple grain to sorghum will require strong anthropological intervention than a mere scenario on water saving techniques.

The introduction of irrigation is not only the most expensive scenario but also one with high potential negative effects on the river ecosystem. A regulated flow regime will have impacts both in terms of alterations to the flow regime as well as disturbance of the natural balance of the fauna and flora which have benefited from the natural cycles of low and high flows. Unfortunately this scenario has high visibility and would gain acceptance and support across the stakeholders' spectrum (policy makers and beneficiaries).Table 11-4 summarises the impacts of the land use scenarios on water hydrology and water resources.

Table 11-4: Impacts of Land use and water management change scenarios on hydrology.

Scenario	Water yields (mm)	Lateral flow (mm)	Shallow Aquifer (mm)	Surface flow (mm)	ET (mm)	PCP (mm)	PET (mm)
Baseline	*233*	*30*	*120*	*83*	*793*	*1155*	*1846*
1976 map	213	32	122	59	813	1155	1846
Irrigation	280	29	153	98	800	1155	1846
Replace_COR N	242	28	122	92	774	1155	1846
Replace_All	282	33	141	108	727	1155	1846
Fertilizer_100	244	30	127	87	780	1155	1846

12. Conclusions

This chapter summarizes the key findings of the research based on the laid out study objectives and research questions. The challenges encountered in the process of achieving the objectives are discussed. It also suggests some recommendations for future research in the Mara River basin and regarding hydrological modeling using the SWAT model.

12.1. Key findings in view of the research objectives

12.1.1 Research objective 1

Assess the trends in the changes in climate, land cover/land use and vegetation variables in the Mara River Basin

A trend analysis for rainfall was performed for six stations located within or in close proximity to the basin. The analysis of historical rainfall data (1962-2008) in the study area are trending towards decreased rainfall. The 48 years rainfall records for Olenguruone and Baraget, in the forested humid climatic zone, show significant declining trends (18 mm/yr and 9 mm/yr respectively) at 95% probability level. The stations in the non-forested sub-humid zones of the study area, Bomet and Tenwek, show non-significant mixed trends, with an insignificant increase for Bomet (1 mm/yr) and an insignificant decrease at Tenwek (2 mm/yr) at 95% probability level. Narok station, located in the semi arid zone of the basin, shows a significant (90% confidence) decrease (4 mm) in historical rainfall. The highest drop in annual rainfall experienced in the study area in the last 50 yrs was 18mm/yr. There was no autocorrelation of the annual rainfall data for five of the six stations, but the Bomet station data show a strong negative autocorrelation. The spatial distribution of the historical change in rainfall indicates a general decrease along the north-east to south-west axis. A seasonality analysis carried out for Narok station – the station with the longest rainfall record (1960-2010)- indicates that there is no significant difference in the occurence of seasons between the first and the second halves of the study period.

The historical trends (1993-2009) in minimum temperatures for the Kericho, Narok and Kisii stations show a significant (90% confidence level) increase. The highest rise in minimum temperature was experienced in Narok, with an increase of up to 0.04^0C/yr. The rise of annual temperature is caused nearly exclusively by a rise of the minimum temperatures, as the increase of the maximum temperature is insignificantly over the last 17 years (1993-2009). Based on the three stations, the spatial analysis of the minimum temperature changes show a west to east increment. Though insignificant the increment in the maximum temperature, is in the south-east to north-west direction.

The indices for vegetation health, including the annual integrated NDVI (inDVI), the seasonal NDVI, the standardised vegetation index (SVI), the vegetation condition index (VCI) and the vegetation productivity indicator(VPI), have a positive increasing trend and skewness towards good and very good health for the last 12 years (1999-2011). Increase in minimum temp may have contributed to better vegetation health due to availability of more heat units.

Multi-temporal images (1986, 1995, and 2006) from Landsat multispectral sensors (MSS, TM and ETM+) were separately classified into thematic maps using both supervised and unsupervised techniques. Post classification techniques, involving the comparison of satellite imagery pixel by pixel, was implemented on the classified images. Land change analysis indicates that between 1976 and 2006, the area under cropland cover has increased in size by 109%. The other land cover types in the area including shrub cover, forest and grassland decreased by 34%, 31% and 4% respectively.

The changes in the basin over three decades were also analysed by considering unclassified Landsat images of 1986, 1995 and 2006). This was performed by simultaneously projecting each of three NDVI dates through the red, green, and blue (RGB) computer display write functions. Change and no change categories were identified by interpreting the RGB-NDVI color composites. A replacement of a vegetation type with one of lower (higher) NDVI value means a loss (or gain) of vegetation biomass. The biggest loss in vegetation biomass was found to have taken place in the 1990s (1995-2006) due to agriculture expansion and excision of government forest land.

Though deficient in phosphorous the soils in the Upper Mara basin are generally of good quality for crop production with crop yields in the basin above the national levels. The concentration for both total phosphorous and total nitrogen remain acceptably low (< 1 mg/l). However, during the wet season the concentrations of TN and TP are beyond the levels for natural systems, suggesting the influence of anthropogenic interference mostly near the urban centres and in agricultural streams.

12.1.2 Research objective 2

Adapt a process based hydrological model to evaluate and predict the response of hydrological processes to changing climatic, land use and water management conditions under past, present and future conditions

A SWAT model was built for the Upper Mara River basin, using local observed data, global databases and remotely sensed data. Observed data included the climatic, stream flow and crop yield data. Global datasets used in the study were the digital elevation maps (DEM), the soil and the land cover maps. Landsat and SPOT_VGT NDVI remotely sensed databases were used to develop the land map. The Upper Mara basin was defined as the area between the Mau escarpments and the point where the Maasai mara game reserve begins.

Simulations were made with the SWAT model, using different land use maps and climatic inputs. Hereby, it was shown that the use of the "dominant type rule" in assigning the land use classes in the creation of land use/land cover maps significantly affects the simulated water balance. It was also shown that the land cover map with clustered land use classes produced a better hydrological balance than when the thematic map was derived using a per pixel classification. Also, the use of interpolated precipitation data for the three rainfall zones resulted in a better model performance than with individual rainfall stations. The individual observed station have various degree of missing data which was filled during the the interpolation process.

The modification of the model databases (crop databases and management files) to suit the Upper Mara watershed was critical in attaining realistic hydrological components. The daily model was satisfactorily calibrated and validated for stream flow using observed flow data for the period 1970 to 1977 at the Lalgorian bridge gauging station (1LA04). The Nash Sutcliffe Efficiency and percent of bias were 0.55 and 8% respectively. For the water balance, the annual water yield was estimated to be 233mm/yr, with baseflow contributing ca. 67%. This compared well with the observed annual flow of 234mm/yr and a baseflow fraction of 71%. The evapotranspiration component for the catchment and for the individual HRU was within the expected ranges of between 600 and 1000 mm/yr, depending on location (semi-arid or humid) and vegetation type

Data from remote sensing and field survey were successfully implemented in the validation of the SWAT model outputs. The timing of the planting season determined from the crop calendar, which was developed during field surveys, agreed with that of the remote sensed LAI. Observed yields and those simulated by the SWAT model were comparable once the model had been adjusted to reflect realistic catchment activities, including farmers' practices of fertilizer use.

12.1.3. Research objective 3

Assess the impacts of climate, land use and water management changes on the sustainable management of water resources in the Mara River Basin.

From a sensitivity analysis of climatic variables it was found that an increase in both minimum and maximum temperatures leads to an increase in the evapotranspiration (ET) and a decrease in the water yield. Higher rainfall input amounts lead to increases in both ET and water yield. Surface heating increase both the surface temperatures and the evaporation so long as there is adequate moisture. The water-holding capacity of the atmosphere increases as temperature rises. With an increase in water carrying capacity of the atmosphere there is enhanced evaporation.

Increase in CO_2 leads to a rise in water yield and a drop in ET. Stomata conductance of plants decline as the atmospheric CO_2 increases, resulting in a reduction of transpiration. With declining transpiration, the total evapotranspiration component of the water balance will decrease with increasing CO_2 build-up. This will lead to a higher water yield in order to maintain a closed water balance. The combination of reduced rainfall and higher temperature make the impact of the drought through increased evapotranspiration.

There is an agreement in all GCMs of higher temperatures in the 2020s (2011-2030) and 2050s (2046-2065) than during the 1980s (1961-1990). For the same periods respectively 80% and 90% of the GCMs agree on an increase in the projected rainfall for the A1B and B1 scenarios of the IPCC. The 2050s are expected to have higher projected rainfall and temperatures than the 2020s. There is a higher variability between the GCMs in the A1B than in the B1 scenarios for both 2020s and 2050s.

The projected impact of the climate change is an increase in the water yield and the evapotranspiration. The analysis of the frequency duration curves of the flows shows   that the maximum flows over 30 and 90 days will increase, but that the 1 day, 3 day and 7 day maxima will decrease. The flow rate for the 100 yrs return flood is expected to increase by 11 and 18% in the 2020 and 2050 A1B scenarios respectively. According to the simulations, there is no significant change to be expected for the minimum flows.

The increase of forest cover to pre 1976 levels will lead to a reduction in the water yield and an increased ET, with possibly a higher transfer of water to the groundwater. The replacement of lowland maize crop - whose production has been seriously affected by the maize lethal necrosis (MLN) disease- with grain sorghum will lead to higher (4%) water yields. Further replacement for all lowland crops with grain sorghum will lead to even higher (21%) water yields. There is a significant reduction (in some HRU up to 50%) in the water stress with the sorghum crop. Use of irrigation leads to both higher river flow and ET, while the average crop yields double in this scenario, even without fertilizer use.

Regarding the impact of the use of fertilisers, it has been shown that the response of the use of DAP fertilizers on the crop yields is soil-dependent. The use of DAP fertilizer significantly and proportionately increases yields for all the soils. The majority of the soil types (86%) would require fertilizer application of between 150 and 220 kg of (18:46:00) DAP/ha to attain the world average maize yields of 3 ton/ha. For two of the soil types in the basin, as little as 50kg/ha would be sufficient to produce the target 3 ton/ha. Excessive use of fertilizer in such fields would not only be a waste of resources but would also pose a risk to the water quality. Other soil types, poorly suited for crop production even with addition of unrealistically high fertilizer amounts, are better off left for natural production systems like grassland.

Whereas 80% of the GCM predict a wetter future, historical trends analysis show an opposite declining direction in rainfall. This disagreement makes it difficult to predict the impacts of climate change on the hydrology. Based on the historical trends, an increase in temperature and decrease in rainfall potent a disastrous scenario since the water yields will decrease and the ET increase. Based on this scenario the adoption of grain sorghum growing in place of and as a rotation for corn and the use of fertilizers should be considered as part of the coping mechanisms to the climate change.

If the GCM projections are upheld and there is an increase rainfall, there will be an increase in the water flow. The increased high flows could be stored in reservoir, and the water used for crop irrigation in the dry period for the semi-arid areas to supplement rainfed agriculture. It was shown that the use of expected excess flows due to climate change for irrigation will lead to increased yields and reduced water stress in the dry periods.

12.2 Contributions of this study

The cluster and averaging (CA) method for interpolating rainfall was introduced in this study. The simple method considers the topography and the rainfall amounts and groups all the stations that cluster together. The mean of the rainfall in the groups is used as the station rainfall for that subbasin. The method was validated by performing kriging of the rainfall data. The CA clusters falls into the same zones as those formed in the kriging process. The method was compared with the inverse distance weighting method and the nearest neighbor methods on SWAT, and found to produced the best NSE.

Although thematic maps have been produced from remote sensed data before, in this study a f crop level map has been developed – using NDVI data - specifically for use as an input map for hydrological modeling. The main difference with the other studies is the limited amount of resources (effort) put into the mapping and the level of details produced. The 1x1 km NDVI map produced a land use map with accurate enough details , which may only be possible to produce with extensive ground truthing field work using even finer resolution 30x30m Landsat maps. The key to the land mapping was the use of the field information especially the crop calendar developed during field surveys. The information of the timing of planting and harvesting was crucial for the separation of spectral signatures mediated by the phenological differences.

The NDVI map was used to improve on the water balance in the modeling process by better representing the processes on the ground instead of using generic land cover types. Classifying land maps to crop level enables the adjustment of the crop and managemement databases to finer details due specificity of crop charactersictics.

Distributed data has been use previously for the validation of hydrological models especially in ungauged catchments. In this study the land unit / soil unit (LUSU) was introduced. The concept is based on identifying the soil type and the land cover at any particular physical point. The SWAT model model can be subdivided into user defined units corresponding to the commonly used local mapping units. The HRU outputs can tehn be clustered to user defined units). This makes it easier to compare the model outputs with measured outputs archived at the maping unit. The LUSU was used to validate the leaf area index in SWAT. Its ease of implementation reduces the dependence on GIS expertise.

12.3 Challenges

The achievement of the objectives of this study was not without challenges. The main challenge in the study area is access to good observed data, both in quality and quantity. The principal custodians for these datasets are state corporations. Due to lethargy, underfunding and understaffing, there was a near total collapse of the monitoring systems, especially for stream flow, in the 1980s and 1990s. For most of the river basins, especially where there are no hydro electric power (HEP) dams, there was little emphasis to monitor the flows. This problem created a large gap in the data availability. This is a big shortcoming in the modeling process since the calibration of a hydrological model with observed data is a key step in adapting the model to the basin conditions.

The analysis of impacts from different scenarios regarding changes due to both the land use and the climate can only be meaningful if the model is well adapted. Efforts by key players, including the Water Resources Management Authority and the Lake Victoria Basin Authority, to improve on the hydrometeorogical network in the region is commendable. The vandalism of these new stations for material re-sale as scrap metal is disturbing and a great disservice both to the local community and the larger scientific world.

Lack of a water quality (including sediments) monitoring scheme is a big hindrance to monitoring land use changes. Some of the best manifestations of degradation of watershed are the sediment loading and the turbidity. The unavailability of an automatic sampling gear to capture the freak storms is a setback towards understanding the magnitude of nutrient fluxes and sediment transport problem.

The attempt to link trends in climatic variables to trends vegetation response in study was unfruitful. The idea was to derive a method that could use vegetation indices to deduce climatic anomalies in the absence of climatic data. The high precipitation in the upper humid zones study area showed there was no extractable correlation. A better correlation could have been expected in the semi-arid zones, but lack of rainfall and temperature data in the lower sections of the basin hindered the successful implementation such an attempt.

The lack of a clear land planning policy caused the subdivision of land into small agricultural production units, which are almost economically unviable. Also, due to the small irregularly shaped land units, remote sensing classification tools produce land cover maps with a very large number of small, fragmented mixed signal land units.

The modelling of the tropical forest land cover types with the SWAT has remained a big challenge. The model was developed especially for use in temperate agricultural watersheds. Attempts to improve on the ET simulation for forest produced mixed results. The initialisation of the leaf area index and the biomass in the model was the most promising. Although the buildup of more biomass in the forest was not attained, there was no characteristic annual drop of the biomass to zero, as experienced in other studies and in the default simulation for this study. The requirement for the initialization of the potential heat unit (PHU_LT) if the vegetation is growing at the start of the simulation (IGRO=1) was a major cause of SWAT simulating unrealistic low forest ET.

12.4 Limitations of SWAT modeling for the Upper Mara basin

The SWAT model has been adapted for the Upper Mara basin. The selection of the Old mara bridge streamflow gauging site presented serious data challenges. The gauging station was selected for this study due to its strategic location. It captures the anthropogenic activities taking place upstream of the key Masai Mara - Serengeti ecosystem. The site was established in the 1970s, and has available records are from 1970 to 1991 albeit with high percentage of missing data. Quarterly discharge measurements at a bridge 50m upstream of the station were recorded, before the station was refitted with an automatic sensor during the -Jan - feb dry season of 2010 (Ngessa, personal communication).

Though the selected period (1970-1977) had better available data with fewer missing points, it might have been short to cover the whole range of internnual climate variablity cycle. However, the use of short data periods is not unique to this study. Dessu and Melese 2012, also used "7 year data segments" for their calibration of the Nyangores and Mara mines.

The use of multisite calibration serves to address the hydrological model to different gauging stations. In normal practices some stations will be poorly simulated than others (Arnold et al., 2012). SWAT hydrological model was initially developed for ungauged or poorly gauged catchments. The use of performance criteria that relies on comparing the observed and simulated values also sets out performance thresholds upon which a minimum performance must be met for the model to be deemed satisfactory for use in impact assessments on water resources.

The poor performance of the Amala and Nyangores gauging station has been reviewed Dessu and Melese, 2012 and Mango et al., 2012. Despite the discharge data available for use in both studies being of seemingly fairly long duration (1978-1992, 1996-2008 respectively), both authors concluded that the poor performance of the SWAT model in this upper section was due to poor data quality.

The changing environment in the Upper Mara may be presenting unique hydrological complications

1. The significant drop in the observed precipitation might be affecting the fitting of the water balance. In a study rating curve in the Nyangores sub-catchment of the study area, Juston et al., 2013, found only "subtle changes in the discharge" which they attributed to the change in the river morphology near the gauging site after the construction of a bridge. These findings indicates that despite the reduction in the input part of the water balance, the measured (monitored) surface flow part of the output shows no change. This could imply that the baseflow part might have changed and gone unnoticed. For the water balance to be realized, a drop in input must be accompanied by a corresponding decrease in the output sections. Since the objective function in the calibration process is based on the fitting the observed surface flow to the flow simulated at a discharge point, the ground water component is not considered in the fitting of the objective function.

2. The destruction, removal and the replacement of dense forest cover with sparse cropland has been shown in controlled catchments to lead to increase in water yield. There is evidence of removal of forest vegetation in the Upper Mara basin in this study which is consistent with other findings (Mati et al., 2008, Olang et al., 2011, Mango et al., 2012).

Coincidentally, the removal of forest in the Upper Mara basin seems to have "conveniently compensated" for the reduction in rainfall. As a result the recorded flow in the river may not have changed significantly even with a significant drop in the precipitation in the basin.

3.The two sub-basins in the Upper Mara basin (Nyangores and Amala) consist of geological structure formed as a result of volcanic activity, this consists of volcanic ashes mainly trachytes and tuffs. The ashes form a deep soil layer with high hydraulic conductivity. There is high possibility of water loss to the deep aquifers. The Upper Mara basin is also located in the Rift Valley. The presence of faults lines may also be contributing to substantial loss of water from the basin, imposing difficulties in the closing of the water balances. In the SWAT simulation the waterbalance does not close 100% indicating that some water is still unaccounted for. The restriction of the amount of water that could be transferred to the deep aquifer using the parameter Rrchr_DP maybe responsible for this anomaly.

The performance of the adpoted SWAT model was assessed under the three classes/categories of model evaluation described by van Griensven et al., 2012. In the "fit-to-observations" class, the study used three statistical parameters with results for daily time step ranging from satisfactory to good. The assessment of model performance with monthly and yearly time steps would have yielded better performance but the daily time step was selected to give more comparison points and to reduce errors in accumulating months and years with missing daily observations. In the "fit-to-reality" class, the water yield and the evapotranspiration in the hydrological balance of the Upper Mara were within ±5% of the expected range of values. The "fit-to-purpose" class saw the model simulate crop yields to fairly mirror the yields produced in the different zones and soil types. The adapted SWAT model was also able to match the timing of the growing period of perenial crops as compared with he remote sensed Leaf air index.

Despite the short period of usable observation data, the adapted SWAT model has been proven to not only simulate the physical processes well but also reproduce management activities taking place in the watershed. Though this limits the extent to whcih the results and findings can be applied in water resources management, it will at least trigger debate on the need to set-up and manage hydrometeorological stations.

12.5 Recommendations

Based on the experience gained in this study, the following recommendations touching on the model inputs, model routines and model implementation are offered:

The wealth of data currently available from remote sensing should be aptly applied to monitor hydrological processes. With the availability of near-real time remote sensed ET, LAI, NDVI and climate data for all locations in the globe (albeit at coarse resolution), a dynamic feedback mechanism could be established. Realistic modeling of hydrological processes may be improved if the land cover input is made dynamicFurther, the SWAT simulator processes could also be improved by hard-wiring default operations for known perenial landcover types to automatically and realistically simulate the biomass accumulation processes. The SWAT crop database needs to be expanded to include information on major plant communities and types in tropical regions. Such information specific for tropical climates include: leaf area index, timing and percentage of senescence and base temperature for growth.

The use of hydrological models in data scarce and data poor catchments is a big challenge. Methods of calibration and validation of these models using non-statistical objective functions and performance criteria need to be evaluated to quantify the skill of models to fit to hydrological processes.

The biotic pump theory has been touted to provide a general physical platform for analysing the critical role of land use and cover changes. The study recommends the testing of the theory to assess if land degradation is responsible for the decline in regional rainfall against projected global increase.

The application of fertilizer has both economic and environmental consequences. Soil fertility testing is not performed routinely except only by the large commercial farming enterprises. The study recommends the development of generic fertilizer application maps which could serve as a quick guide to farmer fertilizer use depending on the locations' soil type. Delineation of exclusion zones for non cultivated agricultural production could also be developed.

A water quality monitoring programme could be established to monitor the impacts of increased agricultural activity. The planned government policy to make Kenya food sufficient through a green revolution involving the use of fertiliser will increase the threat of nutrient loading into the river. The continued use of ad hoc water quality monitoring exercises may serve a short term goal, but a longer term sustainable mechanism should be established to give early warning signals to avert disasters and maintain ecological integrity.

13. References

Abbaspour, K.C., M. Vejdani, and S. Haghighat, 2007. SWAT-CUP calibration and uncertainty programs for SWAT, MODSIM 2007 International Congress on Modelling and Simulation, Modelling and Simulation Society of Australia and New Zealand, 1596-1602.

Abbaspour, K.C., A. Johnson, M. van Genuchten, 2004. Estimating uncertain flow and transport parameters using a sequential uncertainty fitting procedure, Vadose Zone Journal, 3:1340-1352.

Abbott, M., J. Bathurst, J. Cunge, P. O'Connell, and J. Rasmussen, 1986. An introduction to the European Hydrological System Systeme Hydrologique Europeen, ``SHE'', 2: Structure of a physically-based, distributed modelling system, Journal of Hydrology, vol. 87 (1-2), 61-77.

Aber, J.D., S.V. Ollinger, C.A. Federer, P.B. Reich, M.L. Goulden, D.W. Kicklighter, J.M. Melillo, and R.G. Lathrop Jr., 1995. Predicting the effects of climate change on water yield and forest production in the northeastern United States, Climate Research 5, 207–222.

Adams, I.P., D.W. Miano, Z.M. Kinyua, A. Wangai, E. Kimani, N. Phiri, R. Reeder, V. Harju, R. Glover, U. Hany, R. Souza-Richards, P. Deb Natha, T. Nixon, A. Fox, A. Barnes, J. Smith, A. Skelton, R. Thwaites, R. Mumford, and N. Boonham, 2013. Use of next-generation sequencing for the identification and characterization of Maize chlorotic mottle virus and Sugarcane mosaic virus causing maize lethal necrosis in Kenya, Plant Pathology Plant Pathology, 62, 741–749.

Agrawal, S., P.K. Joshi, Y. Shukla, and P.S. Roy, 2003. SPOT VEGETATION multi temporal data for classifying vegetation in south central Asia, Current Science, 84(11), 1440-1448.

Allen, R.G., L.S. Pereira, D. Raes, and M. Smith, 1998. Crop evapotranspiration guidelines for computing crop water requirements, FAO, Rome.

Al-amri, S.S., N.V. Kalyankar, and S.D. Khamitkar, 2010. A comparative study of removal noise from remote sensing image, International Journal of Computer Science Issues, Vol. 7, Issue. 1 (1), 1148.

Amatya, D.M., R.W. Skaggs, and J.D. Gregory, 1995. Comparison of methods for estimating REF-ET, Journal of Irrigation and Drainage Engineering, 121 (6), 427-435.

Ameel, J.J., R.P. Axler, and C.J. Owen, 1993. Persulfate digestion for determination of total nitrogen and phosphorus in low-nutrient waters, American Environmental Laboratory,Volume 10/93.

Anderson, J.R., E.E. Hardy, J.T. Roach, and R.I.E. Witmer, 1976. A Land Use and Land Cover Classification System for Use with Remote Sensor Data, Geological Survey Professional Paper 964.

Anyona, D.N., G.O. Dida, A.V. O. Ofulla, and P.O. Abuom, 2012. Influence of urban centres on nutrient and microbial levels along Amala and Nyangores tributaries of the Mara River, 3rd Lake Victoria Basin Scientific conference, Entebbe, Uganda.

APHA-AWWA-WPCF., 1985. Standard methods for the examination of water and wastewater, 16 Ed. Washington, EUA.

Arel-Bundock V., 2013. Description: Search, extract and format data from the World Bank's World Development Indicators, Version: 2.4.URL: http://www.umich.edu/~varel , https://www.github.com/vincentarelbundock/WDI accessed on Date: 10-22-2013.

Arnold, J.G., and N. Fohrer, 2005. SWAT2000: Current capabilities and research opportunities in applied watershed modelling, Hydrol.Process, 19, 563–572.

Arnold, J.G., P.M. Allen, M. Volk, J.R. Williams, and D.D. Bosch, 2010. Assessment of different representations of spatial variability on SWAT model performance, American society of agricultural and biological engineers, Vol. 53(5), 1433-1443.

Arnold, J.G., R. Srinivasan, R.S. Muttiah, and J.R. Williams, 1998. Large-area hydrologic modeling and assessment: Part I.Model development, J. American Water Resource. Association, 34(1), 73-89

Arnold, J.G., and P.M. Allen, 1993. A comprehensive surface groundwater flow model, J. Hydrol. 142 (1-4), 47-69.

Arnold, J.G., and P.M. Allen, 1999. Automated methods for estimating baseflow and groundwater recharge from streamflow records, J. American Water Resource. Association, 35 (2), 411-424.

Arnold, J.G., J.R. Williams, and D.R. Maidment, 1995a. Continuous time water and sedimentrouting model for large basins, J. Hydrol. Eng. ASCE 121(2): 171-183.

Arnold, J. G., P. M. Allen, R. S. Muttiah, and G. Bernhardt, 1995b. Automated base flow separation and recession analysis techniques. Groundwater 33(6): 1010-1018.

Asner, G. P., J.M. O. Scurlock, and J.A. Hicke, 2003. Global synthesis of leaf area index observations: implications for ecological and remote sensing studies, Global Ecology & Biogeography, 12, 191–205.

Assefa, Y., and S.A. Staggenborg, 2010. Grain sorghum yield with hybrid advancement and changes in agronomic practices from 1957 through 2008, Agronomy Journal, 102 (2), 703-706.

Banana, A., M. Buyinza., E. Luoga, and P. Ongugo, 2010. Emerging Local Economic and Social Dynamics Shaping East African Forest Landscapes,chapter 17 in Forests and Society – responding to Global drivers of Change, IUFRO World Series Volume 25 www.iufro.org/download/file/5904/4668/315-334_pdf/

Bannari, A., D. Morin, and F. Bonn, 1995. A review of vegetation indices, Remote Sensing Reviews, 13, 95-120.

Baret, F., O. Hagolle, B. Geiger, P. Bicheron, 2006. Leaf Area Index (LAI) in Bartholomé (edits) VGT4Africa user manual, EUR 22344, 2004

Batjes, N.H., and P. Gicheru. 2004. Soil data derived from SOTER for studies of carbon stocks and change in Kenya, ISRIC World Soil Information Report 2004/01.

Bayarjargal, Y., A. Karnieli, M. Bayasgalan, S. Khudulmur, C. Gandus, and C.J. Tucke, 2006. A comparative study of NOAA–AVHRR derived drought indices using change vector analysis, Remote Sensing of Environment, 105, 9–22.

Beckers, J., B. Smerdon, and M. Wilson, 2009. Review of Hydrologic Models for Forest Management and Climate Change Applications in British Columbia and Alberta, FORREX SERIES 25.

Benjamin, J.R., and C.A. Cornell, 1970. Probability, statistics, and decision for civil engineers, McGraw-Hill, New York.

Bernoux, M., L. Bockel, G. Branca, A. Carro, L. Lipper, and G. Smith, 2010. Ex-ante greenhouse gas balance of agriculture and forestry development programs, Scientia Agricola 67(1), 31-40.

Beven K., and A. Binley, 1992. The future of distributed models model calibration and uncertainty prediction, Hydrological Processes, 6 (3), 279–298.

Bhat, M., D. Ombara, W. Kasanga, M. McClain, and G. Atisa, 2009. Payment for Watershed Services in the Mara River Basin, Florida International University, Miami, FL

Bicknell, B.R., J.C. Imhoff, J.L. Kittle, Jr., T.H. Jobes, and A.S. Donigian, Jr. 2001. Hydrological Simulation Program-Fortran, Version 12: User's Manual, Mountain View, Cal.: AquaTerra Consultants.

Binaghi, E., P.A. Brivio, P. Ghezzi, and A. Rampini, 1999. A fuzzy set-based accuracy assessment of soft classification, Pattern Recognition Letters, 20:935–948.

Blaschke, T., and J. Strobl, 2000. What's wrong with pixels? Some recent developments interfacing remote sensing and GIS, GIS 6/01, 12-17.

Blasone, R.S., H. Madsen, and D. Rosbjerg, 2007. Parameter estimation in distributed hydrological modelling:comparison of global and local optimisation techniques, Nordic Hydrology Vol 38 (4–5), 451–476.

Boisrame, G., 2011. WGNmaker.xlsm usermanual, http://swat.tamu.edu/ media/ 41586/ wgen-excel.pdf.

Boko, M., I. Niang, A. Nyong, C. Vogel, A. Githeko, M. Medany, B. Osman-Elasha, R. Tabo and P. Yanda, 2007. Africa: Climate Change 2007: Impacts, Adaptation and Vulnerability. Contribution of Working Group II to the Fourth Assessment Report of the Intergovernmental Panel on Climate Change, M.L. Parry, O.F. Canziani, J.P. Palutikof, P.J. van der Linden and C.E. Hanson, Eds., Cambridge University Press,Cambridge UK, 433-467.

Borah, D.K., and M. Bera, 2003. Watershed-scale hydrologic and nonpoint-source pollution models: Review of applications, Trans. ASAE 47(3): 789-803.

Borah D. K., J. G. Arnold, M. Bera, E. C. Krug, and X.Z. Liang, 2007. Storm Event and continuous hydrologic modeling for comprehensive and efficient watershed simulations, J. Hydrologic Engineering. Volume 12 (6), 605-616.

Bosch, D., F. Theurer, R. Bingner, G. Felton, and I. Chaubey, 1998. Evaluation of the AnnAGNPS Water Quality Model, SDA-CSREES Southern Region Research Project S-273.

Bouraoui, F., and T. Dillaha, 1996. ANSWERS-2000: Runoff and sediment transport model, J. Environ. Eng. 122 (6): 493-502.

Breuer, L., K. Eckhardt, and H.G Frede, 2003. Plant parameter valued for models in temperate climates, Ecological Modelling, 169, 237-293.

Brown, L.C., and T.Barnwell, 1987. The Enhanced Stream Water Quality Models QUALE-2E and QUALE-2E-UNCAS: Documentation and user manual, Environmental Research Laboratory US EPA. Athens, Georgia, USA.

Bruijnzeel, L.A, and J. Proctor, 1995. Hydrology and biogeochemistry of tropical montane cloud forests: what do we really know?, in L.S Hamilton et al. (eds), Tropical montane cloud forests, Springer-verlag, New York.

Bruijnzeel, L.A., 1992. Managing tropical forest watersheds for production: where contradictory theory and practice co-exist. In:Miller, F.R., Adam, K.L. (Eds.), Wise Management of Tropical Forests, Oxford Forestry Institute, Oxford, 37–75.

Burn D.H., and M.A.H Elnur, 2002. Detection of hydrological trends and variability, J. of Hydrol. 255, 107-122.

Burrough, P., and R. McDonnell, 1998. Principles of Geographic Information Systems, Oxford University Press, New York, NY, 333 pp.

Campbell, J.B., 2002. Introduction to remote sensing, The Guilford Press: New York.

Carter, M.R., and E.G. Gregorich, (Eds), 2007. Soil Sampling and Methods of Analysis. CRC Press, Inc. Boca Raton, FL, USA.

Cess R.D., G.L. Potter, J.P. Blanchet, G.J. Boer, S.J. Ghan, J. Hansen, J.T. Kiehl, H. Le Treut, X.Z. Li, X.Z. Liang, B.J. McAvaney, V.P. Meleshko, J.F.B. Mitchell, J.J. Morcrette, D.A. Randall, L.J. Rikus, E. Roeckner, U. Schlese, D.A. Sheinin, A. Slingo, A.P. Sokolov, K.E. Taylor, W.M. Washington, R.T. Wetherald, and I. Yagaii, 1990. Intercomparison and interpretation of cloud-climate feedback processes in nineteen atmospheric general circulation models, J Geophys Res 95 (16), 601–615.

Chaplot, V., 2007. Water and soil resources response to rising levels of atmospheric CO_2 concentration and to changes in precipitation and air temperature, Journal of Hydrology. 337(1-2), 159-171.

Chaubey I., A. S. Cotter, T. A. Costello, and T. S. Soerens, 2005. Effect of DEM data resolution on SWAT output uncertainty, Hydrological Processes,Volume 19 (3), 621–628.

Chen J., F.P. Brissette, and R. Leconte, 2011, Uncertainty of downscaling method in quantifying the impact of climate change on hydrology, Journal of Hydrology 401, 190-202.

Chen J., P. Jonsson, M. Tamura, Z. Gu, B. Matsushita, and L. Eklundh, 2004. A simple method for reconstructing a high-quality NDVI time-series data set based on the Savitzky–Golay filter, Remote Sensing of Environment, 91, 332–344.

Cho, S.M., and M.W. Lee, 2001. Sensitivity considerations when modeling hydrologic processes with digital elevation model, Journal of the American Water Resources Association 37, 931–934

Chua, T.H., 2012. Modelling the effect of temperature change on the extrinsic incubation period and reproductive number of Plasmodium falciparum in Malaysia. Tropical Biomedicine 29 (1), 121–128.

CIMYT., 2012. Maize lethal necrosis (MLN) disease in Kenya and Tanzania: Facts and actions, accessed 06/06/2013.

Congalton, R. G., 1991. A review of assessing the accuracy of classifications of remotely sensed data. Remote sensing environment. 37:35-46.

Congalton, R.G., 2005. Thematic and Positional Accuracy Assessment of Digital Remotely Sensed Data, Proceedings of the Seventh Annual Forest Inventory and Analysis Symposium.

Congalton, R.G., and K. Green, 1993. A practical look at the sources of confusion in error matrix generation. Photogrammetric Engineering and Remote Sensing, 59, 641–644.

Costa, M.H., A. Botta, and J.A. Cardille, 2003. Effects of large-scale changes in land cover on the discharge of the Tocantins River, Southeastern Amazonia, Journal of Hydrology 283, 206–217.

Cotter, A.S., I. Chaubey, T.A. Costello, T.S. Soerens, and M.A. Nelson, 2003. Water quality model output uncertainty as affected by spatial resolution of input data. J. of the American Water Resources Association (JAWRA) 39(4), 977-986.

Cronshey, R.G., and F.D. Theurer, 1998. AnnAGNPS—Non-Point Pollutant Loading Model, Proceedings First Federal Interagency Hydrologic Modeling Conference, 19-23 Las Vegas, NV, 9-16.

Curran, P.J., 1985. Principles of Remote Sensing, Longman, New York, 282 pp.

Daniel, E.B., J.V. Camp, E.J. LeBoeuf, J.R. Penrod, J.P. Dobbins, and M.D. Abkowitz, 2011. Watershed Modeling and its Applications: A State-of-the-Art Review, The Open Hydrology Journal, 5, 26-50.

Davenport, M.L., and S.E. Nicholson, 1993. On the relation between rainfall and the Normalized Difference Vegetation index for diverse vegetation types in East Africa, International journal Remote Sensing 14(12), 2369-2389.

Davis, J.C., 1986.Statistics and Data Analysis in Geology, John Wiley & Sons, New York

De Bie, C.A.J.M., 2002. Novel approaches to use RS-products for mapping and studying agricultural land use systems. In: ISPRS 2002 TC-VII : Commission VII, Working Group VII-2.1 on sustainable agriculture, International symposium on resource and environmental monitoring, Hyderabad. 9p. http://www.itc.nl/library/Papers/0019.pdf Last accessed 27 Dec 2012

Dechmi, F., J. Burguete, and A. Skhiri, 2012. SWAT application in intensive irrigation systems: Model modification, calibration and validation, Journal of Hydrology, vol.470-471, 227-238.

Dessu, S.B., and A.M. Melesse, 2012. Impact and uncertainties of climate change on the hydrology of the Mara River basin, Kenya/Tanzania, Hydrol.Process. DOI:10.1002/hyp.9434.

de Sherbinin, A., 2002. Land-Use and Land-Cover Change, A CIESIN Thematic Guide. Palisades, NY: Center for International Earth Science Information Network of Columbia University. Available on-line at http://sedac.ciesin.columbia.edu/tg/guide_main.jsp.

de Wit, M., and J. Stankiewicz, 2006. Changes in surface water supply across Africa with predicted climate change, Science 311 (5769) 1917-1921.

Discoe, P., 2009. VT buider, VT project

DHI, 1993. MIKE-SHE, Technical Reference Manual – Water movement module – release 1.0. Danish Hydraulic Institute, Denmark, 136 pp.

Di Gregorio A., and L.J.M. Jansen, 2000. Lands cover classification system (LCCS), FAO.

Di Gregorio, A., and L.J.M. Jansen, 1997. Part I-Technical document on the Africover Land Cover Classification Scheme, in: FAO. Africover Land Cover Classification. FAO, 4-33.

Di Luzio, M., R. Srinivasan, and J.G. Arnold, 2004. A GIS-Coupled Hydrological Model System for the Watershed Assessment of Agricultural Nonpoint and Point Sources of pollution, Transactions in GIS Volume 8 (1),113–136.

Dixon, J., A. Gulliver, and D. Gibbon, 2001. Farming systems and poverty. Improving farmers' livelihoods in a changing world. FAO (Food and Agriculture Organization), Rome, and World Bank, Washington, DC.

Dixon, J., A. Tanyeri-Abur, and H. Wattenbach, 2003. Context and Framework for Approaches to Assessing the Impact of Globalization on Smallholders", In Dixon J., K. Taniguchi and H. Wattenbach (eds), Approaches to Assessing the Impact of Globalization on African Smallholders: Household and Village Economy Modeling, Proceedings of Working Session Globalization and the African Smallholder Study, FAO and World Bank, Rome, Italy, Food and Agricultural Organization United Nations.

Doherty, J., 2005. PEST: Model Independent Parameter Estimation, User Manual (5th edn), Watermark Numerical Computing, Brisbane, Australia.

Donatelli M., G. Bellocchi, and L. Carlini, 2006. Sharing knowledge via software components, Models on reference evapotranspiration, European Journal of Agronomy 24:186–192

Douglas-Mankin, K.R., R. Srinivasan, and J.G. Arnold, 2010. Soil and Water Assessment Tool: current developments and applications, Transactions of the ASABE, Vol. 53 5), 1423-1431.

Downer, C.W., and F.L. Ogden, 2006. Gridded Surface Subsurface Hydrologic Analysis (GSSHA) User's Manual Version 1.43 for Watershed Modeling System 6.1, System-Wide Water Resources Program ERDC/CHL SR-06-1.

Downing C., F. Preston, D. Parusheva, L. Horrocks, O. Edberg, F. Samazzi, R. Washington, M. Muteti, P. Watkiss, and W. Nyangena, 2008; Final Report, Kenya, Climate Screening and Information Exchange ED05603-Issue 2.

Droogers, P., and R.G. Allen, 2002. Estimating Reference Evapotranspiration under inaccurate data conditions, Irrigation and Drainage Systems,16 (1), 33–45.

Droogers, P., S. Kauffman, K. Dijkshoorn, W. Immerzeel, J. Huting, S. Mantel, and G. van Lynden, 2007. Green and Blue water Services in Tana River Basin, Kenya. Exploring options using integrated modeling frame work, Green Water Credits Report 3, ISRIC Report 2007.

Droogers P., and J. Aerts, 2005. Adaptation strategies to climate change and climate variability: A comparative study between seven contrasting river basins, Physics and Chemistry of the Earth 30, 339–346.

Druilhe, Z., and J. Barreiro-Hurlé, 2012. Fertilizer subsidies in sub-Saharan Africa, ESA Working paper No. 12-04, FAO, Rome.

Duda, R.O., P.E. Hart, and D.G. Stork, 2001. Pattern Classification. New York, John Wiley & Sons.

Duflo, E., M. Kremer, and J. Robinson, 2008. How High are Rates of Return to Fertilizer? Evidence from Field Experiments in Kenya, American Economics Association Meetings, New Orleans LA.

Durbin, J., and G. S. Watson, 1951. Testing for Serial Correlation in Least Squares Regression, II .Biometrika 38 (1–2): 159–179.

Easterling, W.E., N.J. Rosenberg, M.S. McKenney, A. Jones, C. Dyke, P.T. J.R Williams, 1992. Preparing the erosion productivity impact calculator (EPIC) model to simulate crop response to climate change and the direct effects of CO_2, Agricultural and Forest Meteorology 59, 17–34.

Ehret, U., and E. Zehe, 2010. Series distance - an intuitive metric for hydrograph comparison, Hydrological Earth system sciences Discussion paper, 7, 8387-8425.

Eklundh, L., and P. Jönsson, 2010. TIMESAT 3.0, Software Manual.

Ellis, E., 2010. Land-use, In Encyclopedia of Earth. Eds. Cutler J. Cleveland, Washington, D.C.

Ellis E., 2009. Land-cover, In Encyclopedia of Earth. Eds. Cutler J. Cleveland, Washington, D.C.

ERDAS, 2005. Field Guide, Leica Geosystems Geospatial Imaging, LLC.

FAO/UNESCO/ISRIC, 1988. Soil Map of the World. Revised Legend. World Soil Resources Report 60. (FAO: Rome).

FAO, 2005. AQUASTAT – database, Food and Agriculture organization, Rome. (available at http://www.fao.org/).

FAO, 2005. Land Cover Classification System (LCCS): Classification concepts and user manual. Software version 2, Food and Agriculture organization, Rome. 190p.

FAO, 2006. Fertilizer use by crop, Fertilizer and plant nutrition bulletin 17, Food and Agriculture organization, Rome.

FAO, 2013. CropWater Information: Maize FAO- Water Development and Management Unit. Food and Agriculture organization, Rome, accessed on 12/06/2012 http://www.fao.org/nr/water/cropinfo_maize.html.

FAO, 2010., Crop calendar - An information tool for seed security, Food and Agriculture organization, Rome. http://www.fao.org/agriculture/seed/cropcalendar/welcome.do.

FAO, 1988,.Soils map of the world: revised legend. Food and Agriculture Organization of the United Nations, Rome. 119 p.

Farr, T.G., and M. Kobrick, 2000. Shuttle Radar Topography Mission produces a wealth of data, Amer. Geophys. Union Eos, v. 81, 583-585.

Fermont, A., and T. Benson, 2011. Estimating Yield of Food Crops Grown by Smallholder Farmers A Review in the Uganda Context, International Food Policy Research Institute (IFPRI) Discussion Paper 01097.

Fick, A.A., C.A. Myrick, and L.J. Hansen, 2005. Potential impacts of global climate change on freshwater fisheries. A report for WWF, Gland, Switzerland.

Ficklin, D.L., Y. Luo, E. Luedeling, and M. Zhang, 2009. Climate change sensitivity assessment of a highly agricultural watershed using SWAT, Journal of Hydrology 374, 16–29.

Fontaine, T.A., J.F. Klassen, R.H.Hotchkiss, 2001. Hydrological response to climate change in the Black Hills of South Dakota, USA. Hydrological Sciences 46 (1), 27–40.

Foody, G.M., 2002. Status of land cover classification accuracy assessment, Remote Sensing of Environment 80, 185–201.

Franklin, S.E., 2001. Remote Sensing for Sustainable Forest Management, Lewis Publishers, Boca Raton, FL, 407 pp.

Funk, C., G. Eilerts, F. Davenport, and J. Michaelsen, 2010. A climate trend analysis of Kenya, USGS Open-File Report.

Gao Y., and D. Long, 2008. Intercomparison of remote sensing-based models for estimation of evapotranspiration and accuracy assessment based on SWAT, Hydrological Processes. 22 (25), 4850-4869

Gann.D., 2006. Rainfall estimation raster conversion, data preparation for hydrological modeling with SWAT: Hydrological modeling with SWAT Mara Basin (Kenya/Tanzania).

Gassman P.W., M.R. Reyes, C.H. Green, and J.G. Arnold, 2007. The Soil and Water Assessment Tool: Historical development, applications and future directions, Transactions of the ASABE, Vol. 50(4), 1211-1250.

Gbetibouo, G.A., 2009. Understanding farmers' perceptions and adaptations to climate change and variability: The Case of the Limpopo Basin, South Africa, IFPRI Discussion Paper.

George, H., and M. Petri, 2006. The rapid characterization and mapping of agricultural land-use: A methodological framework approach for the LADA project, FAO, Rome.

Georgakakos K.P., N.E. Graham, T.M. Carpenter, E. Shamir, J. Wang, J.A. Sperfslage, and S. Taylor, 2006. Integrated Forecast and Reservoir Management (INFORM) for Northern California: System Development and Initial Demonstration, HRC Technical Report No. 5.

Gereta, E.J., E. Wolanski, and E.A.T. Chiombola, 2003. Assessment of the environmental, social and economic impacts on the Serengeti ecosystem of the developments in the Mara river catchments in Kenya. Frankfurt Zoological Society.

Gesch, D.B., 2012. Global digital elevation model development from satellite remote-sensing data. in X. Yang and J. Li (edits) Advances in Mapping from Remote Sensor Imagery Techniques and Applications, CRC Press, 91–118.

Gessner, U., M. Niklaus, C. Kuenzer, and S. Dech, 2013. Intercomparison of leaf area index products for a gradient of sub-humid to arid environments in West Africa, Remote Sensing, 5, 1235-1257.

Gibbs R.J., 1970. Mechanisms controlling world water chemistry, Science, New Series, Vol. 170 (3962), 1088-1090.

Gichana, Z. M., M. Njiru, and P. Raburu, 2012. Influence of land use changes on water quality and Benthic biota along Nyangores stream, Mara river basin, Kenya. 3rd Lake Victoria Basin Scientific conference, Entebbe, Uganda.

Githeko, A.K., and W. Ndegwa, 2001. Predicting malaria epidemics in the Kenyan Highlands using climate data: a tool for decision-makers. Global Change and Human Health, 2, 54-63.

Githui, F., W. Gitau, F. Mutua, and W. Bauwens, 2009a. Estimating the impacts of land-cover change on runoff using the soil and water assessment tool (SWAT): case study of Nzoia catchment, Kenya. Hydrological Sciences Journal 54(5), 899- 908.

Githui, F., W. Gitau, F. Mutua, and W. Bauwens. 2009b. Climate change impact on SWAT simulated streamflow in western Kenya. Int. J. Climatol 29: 1823–1834.

Glavan, M., S. White, and I.P. Holman, 2011. Evaluation of river water quality simulations at a daily time step – Experience with SWAT in the Axe catchment, UK, Clean – Soil, Air, Water, 39 (1), 43–54.

Global Water for Sustainability Program (GLOWS), 2007. Water quality baseline assessment report: Mara River basin, Kenya-Tanzania., Florida International University. 61p.

Government of the Republic of Kenya (GOK), 2007. Kenya Vision 2030, Government printer

Government of the Republic of Kenya (GOK)., 2008. First medium term plan, 2008 – 2012, Kenya vision 2030, Globally Competitive and Prosperous Kenya, Government printer.

Government of the Republic of Kenya (GOK)., 2009. Rehabilitation of the Mau Forest Ecosystem, A Project concept prepared by the interim coordinating secretariat, Office of the Prime Minister, on behalf of the Government of Kenya, Government printer.

Goodrich, D.C, C.L. Unkrich, and R.E. Smithand-Woolhiser, 2002. KINEROS2 – A distributed kinematic runoff and erosion model, Proc. 2nd Federal Interagency Conf. on Hydrologic Modeling, July 29-Aug. 1, Las Vegas, NV.

Green.I.R.A., and D. Stephenson, 1986. Criteria for comparison of single event models, Hydrological Sciences Journal, 31 (3), 395—411.

Grimes D.I.F, and E. Pardo-Igúzquiza, 2010. Geostatistical analysis of rainfall, Geographical Analysis, Vol. 42 (2), 136–160.

Guershman, J.P., J.M. Paruelo, and I. C.Burke, 2003. Land use impacts on the normalised difference vegetation index in temperate Argentina, Ecological Applications, 13 (3), 616-628.

Haguma, D., 2007. Development of a hydrologic model of Kagera River basin using remote sensing data, UNESCO-IHE Master Thesis.

Hamed, K.H., A.R Rao, 1998. A modified Mann-Kendall trend test for autocorrelated data, J. Hydrol., 204, pp. 182–196.

Hargreaves, G.H., and Z.A. Samani, 1985. Reference crop evapotranspiration from temperature, Appl.Eng.Agric., 1 (2), 96–99.

Hashemi S. A., 2011. Investigation of relationship between rainfall and vegetation Index by using NOAA/AVHRR satellite images, World Applied Sciences Journal, 14 (11), 1678-1682.

Haughton, N., 2012. Climate model ensemble generation and model dependence, University of New South Wales, Msc Thesis

Hesslerova, P., and J. Pokorny, 2010. Forest clearing, water loss, and land surface heating as development costs, International Journal of Water, 5 (4), 401-418.

Hay, S. I., J. Cox, D.J. Rogers, S.E. Randolph, D.I. Stern, G.D. Shanks, M.F. Myers, and R.W. Snow, 2002. Climate change and the resurgence of malaria in the East African highlands, Nature 415, 905-909.

He X., H. Guan, X. Zhanga, and C.T. Simmons, 2013. A wavelet-based multiple linear regression model for forecasting monthly rainfall, International Journal of Climatology.

Hoffman, C.M., 2007. Geospatial Mapping and Analysis of Water Availability-Demand-Use within the Mara River Basin, , Florida International University, M.S. Thesis.

Holmes Jr, R.R., and K. Dinicola, 2010. 100-Year flood–it's all about chance: U.S. Geological Survey General Information Product 106.

Horne, R., 2009. 3DEM Software for Terrain Visualization and Flyby Animation, Version 20

Horning, N., J. Robison, E.J. Sterling, W. Turner, and S. Spector, 2010. Remote Sensing for Ecology and Conservation, A handbook of Techniques, Techniques in Ecology & conservation Series, Oxford University press, 467pp.

Houghton, R. A., 2005. Above ground forest biomass and the global carbon balance, Global Change Biology 11, 945–958.

Hulme, M., R. Doherty, T. Ngara. et al, 2001. African climate change: 1900–2100, Climate Research 17, 145–168,

ICPAC, 2007. Climate Change and Human Development in Africa: Assessing the Risks and Vulnerability of Climate Change in Kenya, Malawi and Ethiopia 2007/8, Intergovernmental Authority on Development (IGAD) Climate Prediction and Applications Centre (ICPAC).

International Fertilizer Development Center (IFDC), 2012. Kenya Fertilizer Assessment, Muscle Shoals, Alabama, USA, www.ifdc.org.

International Fertilizer Industry Association (IFA), 1992. World fertilizer use manual.Paris. 632 pp.

IPCC, 2007. Climate Change 2007: The Physical Science Basis, Solomon, S., D. Qin, M. Manning, Z. Chen, M. Marquis, K.B. Averyt, M. Tignor and H.L. Miller (eds.). Cambridge University Press, Cambridge, United Kingdom and New York, NY, USA, 996 pp

IPCC, 2001. Climate Change 2001: The Scientific Basis. Contribution of Working Group I to the Third Assessment Report of theIntergovernmental Panel on Climate Change [Houghton, J.T., Y. Ding, D.J. Griggs, M. Noguer, P.J. van der Linden, X. Dai, K.Maskell, and C.A. Johnson (eds.)]. Cambridge University Press, Cambridge, United Kingdom and New York, NY, USA, 881pp.

IPCC, 1994. IPCC Technical Guidelines for Assessing Climate Change Impacts and Adaptations.Prepared by Working Group II [Carter, T.R., M.L. Parry, H. Harasawa, and S. Nishioka (eds.)] and WMO/UNEP. CGER-IO15-'94. University College -London, UK and Center for Global Environmental Research, National Institute for Environmental Studies, Tsukuba, Japan, 59

Jabro J.D., 1992. Estimation of saturated hydraulic conductivity of soils from particle distribution and bulk density data. Trans. ASEA 35 (2), 557-560.

Jacobs, J.H., and R. Srinivasan, 2005. Application of SWAT in developing countries using readily available data. (3rd International SWAT Conference, July 11-15, Zurich).

Jaetzold, R., and H. Schmidt, 1983. Farm management handbook of Kenya. Ministry of Agriculture, Nairobi.

Jaetzold, R., H. Schmidt, Z.B Hornet, and C.A. Shisanya, 2006. Farm management handbook of Kenya. Natural conditions and farm information (Central Province)., Vol 11/C, 2nd edn. Ministry of agriculture/GTZ, Nairobi

Jain A.K., M. N. Murty, and P. J. Flynn, 1999. Data Clustering: A Review, ACM Computing Surveys, 31 (3), 264-323.

Jama, M., 1991. Forest utilisation by people living adjacent to West Mau, Southwestern Mau and Transmara forest reserves, KIFCON Report, Nairobi.

Jang, G.S., K.A. Sudduth, E.J. Sadler, and R.N. Lerch, 2009. Watershed-scale crop type classification using seasonal trends in remote sensing derived vegetation indices, Transactions of the ASABE, 52(5), 1535-1544.

Jarviel, H.P., J.A. Withers, and C. Neal, 2002. Review of robust measurement of phosphorus in river water: Sampling, storage, fractionation and sensitivity, Hydrology and Earth System Sciences, 6 (1), 113–132.

Jayakrishnan, R., R. Srinivasan, C. Santhi, and J.G. Arnold, 2005. Advances in the application of the SWAT model for water resources management, Hydrol. Process. 19, 749–762.

Jensen, J. R., 1986. Introductory Digital Image Processing, Prentice-Hall, Englewood Cliffs, NJ, 227 pp.

Jensen, J.R., 2005. Digital Image Processing: a Remote Sensing Perspective (3rd ed.). Prentice Hall.

Jensen, M.E., R.D. Burman, and R.G. Allen, 1990. Evapotranspiration and irrigation water requirements, ASCE Manuals and Reports on Engineering Practice No. 70, 360.

Johnson M.V., J.D. MacDonald, J.R. Kiniry and J. Arnold, 2009. ALMANAC: A Potential tool for simulating Agroforestry yields and improving SWAT simulation for Agroforestry watersheds, International Agricultural Engineering Journal 18 (1-2), 51-58.

Kamthonkiat, D., K. Honda, H. Turral, N.K. Tripathi, and V. Wuwongse, 2005. Discrimination of irrigated and rainfed rice in a tropical agricultural system using SPOT VEGETATION NDVI and rainfall data, International Journal of Remote Sensing Vol.26 (12), 2527–2547.

Kees, K., and G. N. Ramankutty, 2001. Land Use Changes During the Past 300 Year, Land use land cover and soil sciences, Vol. I.

Kennedy, J., and R.C. Eberhart, 1995. Particle swarm optimization. In Proceedings of the IEEE international conference on neural networks IV Piscataway, 1942–1948.

Kenya National Bureau of Statistics (KNBS), 2011. Kenya 2009, PoPulation and housing census, http://www.knbs.or.ke/docs/KNBSBrochure.pdf

Kiage, L.M., K.B. Liu, N.D. Walker, N. Lam, and O. K. Huh, 2007. Recent land-cover/use change associated with land degradation in the Lake Baringo catchment, Kenya, East Africa: evidence from Landsat TM and ETM+, International Journal of Remote Sensing,Vol. 28, No. 19, 4285–4309.

Kilonzo, F., and J. Obando, 2012. Food production systems and the farmers' adaptive capacity to Climate Change in the Upper Mara River Basin, 3rd Lake Victoria Basin Scientific conference, Entebbe, Uganda.

Kilonzo F., A. Van Griensven, P. Lens, J. Obando, W. Bauwens, 2012. Improving SWAT model performance in poorly gauged catchments, Centre International Capacity Development (CICD) Series, (ISSN 1868-8578).

Kim, N. W., M. Chung, Y.S. Won, and J.G. Arnold, 2008. Development and application of the integrated SWAT–MODFLOW model, Journal of Hydrology, 356, 1–16.

Kimaiyo, J. T., 2004. Ogiek Land Cases and Historical Injustices 1902 – 2004 "Your Resources, Our Relations", Ogiek Welfare Council, Egerton, Nakuru Vol. 1.

Kingston D. G., and R. G. Taylor, 2010. Sources of uncertainty in climate change impacts on river discharge and groundwater in a headwater catchment of the Upper Nile Basin, Uganda, Hydrol. Earth Syst. Sci., 14, 1297–1308.

Kingston, D.G., M.C. Todd, R.G. Taylor, J.R. Thompson, and N.W. Arnell, 2009, Uncertainty in the estimation of potential evapotranspiration Geophysical Research Letters, Vol. 36, Issue 20.

Kiniry, J.R, J.D MacDonald, A.R. Kemanian, B. Watson, G. Putz, and E.E. Prepas, 2008. Plant growth simulation for landscape-scale hydrological modeling, Hydrological sciences journal, Vol. 53, 1030-1042.

Kiniry, J.R., J.R Williams, P.W. Gassman, and P. Debaeke, 1992. A general, process-oriented model for two competing plant species. Trans. ASAE 35 (3): 801-810.

Kinyanjui M.J., 2010. Case study on changes in NDVI values in the period 1998 – 2009: the case of Maasai Mau, Southwestern Mau and Eastern Mau, CASE STUDY 8, Endeleo project.

Kizza, M., A. Rodhe, and C. Xu, 2009. Temporal rainfall variability in the the Lake Victoria basin in the twentieth century. Theoretical and Applied Climatology. Volume 98 (1-2), 119-135.

Klein, D., and J. Roehrig, 2006. How does vegetation respond to rainfall variability in a semi humid West African in comparison to semi arid east African environment?, Center for Remote Sensing of Land Surfaces, Bonn, 28-30.

Klemes, V., 1987. One hundred years of applied storage reservoir theory. Water Resources Management 1 (3), 159-175.

Knight, J. F., and R.S. Lunetta, 2003. An experimental assessment of minimum mapping unit size. IEEE Transactions on geosciences and remote sensing, vol. 41 (9).

Kogan, F.N., 1995. Application of vegetation index and brightness temperature for drought detection, Adv. Space Research, 11, 91-100

Kogan, F. N., 1990. Remote sensing of weather impacts on vegetation, International Journal of Remote Sensing, 11, 1405-1419.

Kosky, K.M., and B.A. Engel, 1997. Evaluation of three distributed parameter hydrologic/water quality models. ASAE Paper No. 97-2010.

Koutsouris, A. J., G. Destouni, J. Jarsjö, and S. W. Lyon, 2010. Hydro-climatic trends and water resource management implications based on multi-scale data for the Lake Victoria region, Kenya, Environmental Research Letters, Volume 5 (3).

Krhoda, G. O., 2001. Preliminary Phase: Project Development and Stakeholder Analysis – The Hydrology of the Mara River. WWF Eastern Africa Regional Program Office - Mara River Catchment Basin Initiative.

Krysanova,V., and A.G. Jeffrey, 2009. Advances in ecohydrological modeling with SWAT-a review. Hydrological sciences journal. 53 (5), 939-947.

Kuczera, G., and E. Parent, 1998. Monte Carlo assessment of parameter uncertainty in conceptual catchment models: the Metropolis algorithm. Journal of Hydrology 211 (1–4), 69–85.

Kundzewicz, Z.W., D. Graczyk, T. Maurer, I. Piskwar, M. Radziejewski, C. Svensson, and M. Szwed, 2005. Trend detection in river flow series: Annual maximum flow, Hydrological Sciences Journal, 50:5, 797-810.

Labatt, C.K., C.M. M'Erimba, and N. Kitaka, 2012. Determining ecological integrity of the Nyangores river using a macro-invertebrate index of biotic integrity (M-IBI), 3rd Lake Victoria Basin Scientific conference, Entebbe, Uganda.

Laflen J.M., L.J. Lane, and G.R. Foster, 1991. WEPP—a next generation of erosion prediction technology, J. Soil Water Conservation, 46, 34–38.

Lakshmi, V., 2004. The role of remote sensing in prediction of ungauged basins, Hydrological processes, Vol. 18 (5), 1029-1034.

Lambin, E.F., B.L. Turner, H.J. Geist, S.B. Agbola, A. Angelsen, J.W. Bruce, O.T. Coomes, R.Dirzo, G.Fischer, C.Folke, P.S. George, K. Homewood, J. Imbernon, Rik Leemans, X Li, E.F. Morano, M. Mortimorep, P.S. Ramakrishnan, J.F. Richards, H. Skanes, W.Steffent, G. D. Stone, U. Svedin, T.A. Veldkamp, C. Vogel, and J. Xu, 2001. The causes of land-use and land-cover change: moving beyond the myths, Global Environmental Change, 11, 261–269.

Lambin, E.F., and H. J. Geist, 2006. Land –use and land-cover change: local processes and global impacts, Springer, 222 pp.

Lambin, E.F., H.J. Geist, and E. Lepers, 2003. Dynamics of land –use and land-cover change in tropical regions, Annu. Rev. Environ. Resource, 28:205–41.

Lambin, E.F., M.D.A. Rounsevell, and H.J. Geist, 2000. Are agricultural land-use models able to predict changes in land-use intensity? Agriculture, Ecosystems & Environment 82 (1-3), 321-331.

Landsberg, J.J., and R.H. Waring, 1997. A generalized model of forest productivity using simplified concepts of radiation use efficiency, carbon balance and partitioning. Forest Ecology and Management, 95, 209-228.

Leavesley, G.H., R.W. Lichty, B.M. Troutman, and L.G. Saindon, 1983. Precipitation-runoff modeling system; user's manual, Water-Resources Investigations Report , no. 83-4238, 207 pp.

Legesse D., C. Vallet-Coulomb, and F. Gasse, 2003. Hydrological response of a catchment to climate and land use changes in Tropical Africa: case study South Central Ethiopia, Journal of Hydrology, 275, 67-85.

Lehner, B., 2005. HydroSHEDS, technical documentation, Version 1.0. World Wildlife Fund US, Washington, DC.

Lepers, E., E.F. Lambin, A.C. Janetos, R. DeFries, F. Achard, N. Ramankutty, and R.J. Scholes, 2005. A Synthesis of Information on Rapid Land-cover Change for the Period 1981–2000, BioScience, Vol. 55 (2), 115-124.

Lettenmaier, D.P., A.W. Wood, R.N Palmer, E.F. Wood, and E.Z. Stakhiv, 1999. Water resources implications of global warming, A U.S. Regional Perspective, Climate Change 43, 537–579.

Lieth C.E., 1973. The standard error of time-average estimates of climatic means, J. Appl. Meteor., 12, 1066-1069

Lillesand, T.M., and R.W. Kiefer, 1987. Remote Sensing and Image Interpretation Somerset, New Jersey, John Wiley and Sons, Inc.

Lim K. J., B.A. Engel, Z. Tang, J. Choi, K. Kim, S. Muthukrishnan, and D. Tripathy, 2005. Automated web GIS based hydrograph analysis tool, WHAT, American Water Resources Association, 41 (6), 1407-1416.

Longobardi, A., and P. Villani, 2010. Trend analysis of annual and seasonal rainfall time series in a Mediterranean area. International Journal of Climatology, 30 (10), 1538-1546.

Loveland, T.R., J.W. Merchant, D. O. Ohlen, and J. F. Brown, 1991. Development of a land-cover characteristics database for the conterminous U.S, Photogrammetric Engineering and Remote Sensing 57 (11), 1453-63.

Luo, Y., C. He, M. Sophocleous, Z. Yin, R. Hongrui, and Z. Ouyang, 2008. Assessment of crop growth and soil water modules in SWAT2000 using extensive field experiment data in an irrigation district of the Yellow River Basin, Journal of Hydrology 352, 139–156.

Lu, J., G. Sun, S. G. McNulty, and D. M. Amatya, 2005. A comparison of six potential evapotranspiration methods for regional use in the southeastern United States, J. American Water Resources Association. 41 (3), 621-633.

Lu, D., P. Mausel, E. Brondízio, and E. Moran, 2004. Change detectiontechniques, International Journal of Remote Sensing, 25 (12), 2365–2407.

Lunetta, R.S., R.G. Congalton, L.K. Fenstermaker, J.R. Jensen, K.C. McGwire, and L.R. Tinney, 1991. Remote sensing and geographic information system data integration:

error sources and research issues, Photogrammetric Engineering & Remote Sensing. 57 (6), 677-687.

LVBC, and WWF-ESARPO, 2010a. Assessing Reserve Flows for the Mara River. Nairobi and Kisumu, Kenya.

LVBC, and WWF-ESARPO, 2010b. Biodiversity Strategy and Action Plan for Sustainable Management of the Mara River Basin. Nairobi and Kisumu, Kenya.

LVBC, USAID, and WWF-ESARPO, 2012. The Mara River Basin SEA: Trans-boundary Strategic Environmental Assessment, Nairobi. LVBC, Kisumu and WWF ESARPO, Nairobi.

MacDonald, J. D., J. R. Kiniry, G. Putz, and E. E. Prepas, 2008. A multi-species, process based vegetation simulation module to simulate successional forest regrowth after forest disturbance in daily time step hydrological transport models, J. Environ. Eng. Sci. 7, 1-18.

Maitima, M.J., S.M. Mugatha, R.S. Reid, L.N. Gachimbi, A. Majule, H. Lyaruu, D. Pomery, S. Mathai, and S. Mugisha, 2009. The linkages between land use change, land degradation and biodiversity across East Africa, African Journal of Environmental Science and Technology Vol. 3 (10), 310-325.

Makarieva, A.M, and V.G. Gorshkov, 2010. The Biotic Pump: Condensation, atmospheric dynamics and climate, International Journal of Water, Volume 5 (4), 365-385.

Mango, L.M., A.M. Melesse, M.E. McClain, D. Gann, and S.G. Setegn, 2011. Land use and climate change impacts on the hydrology of the upper Mara River Basin, Kenya: results of a modeling study to support better resource management. Hydrol. Earth Syst. Sci., 15, 2245–2258.

Manguerra, H.B., and B.A. Engel, 1998. Hydrologic parameterization of watersheds for runoff prediction using SWAT, Journal of the American water resources association, Vol. 34 (5). 1149-1162.

Markula, A., M. Hannan-Jones, and S. Csurhes, 2009. Pest animal risk assessment Red billed quelea (*Quelea quelea*), Queensland government.

Marshall, L., 2005. Bayesian analysis of rainfall-runoff models: Insights to parameter estimation, model comparison and hierarchical model development, PhD Thesis, UNSW.

Masih, I., S. Maskey, S. Uhlenbrook, and V. Smakhtin, 2011. Assessing the impact of areal precipitation input on streamflow simulation using the SWAT model, Journal of the American Water resources association , Vol. 47 (1), 179-195.

Masese, F.O., M.E. McClain, K. Irvine, G.M. Gettel, N. Kitaka, and J. Kipkemboi, 2012. Macro-invertebrate shredders in upland Kenyan streams; influence of land use on abundance, biomass and distribution. 3rd Lake Victoria Basin Scientific conference, Entebbe, Uganda.

Matano, A., D.N. Anyona, G.O. Dida, and A.V.O. Ofulla, 2012. Impacts of Landuse on the physico-chemical properties of water and coliform abundance along the Mara river basin, 3rd Lake Victoria Basin Scientific conference, Entebbe, Uganda.

Mati, B.M., S. Mutie, H. Gadain, P. Home, and F. Mtalo, 2008. Impacts of land-use/cover changes on the hydrology of the transboundary Mara River, Kenya/Tanzania, Lakes & Reservoirs: Research and Management 13, 169–177.

Mayaux, P., E. Bartholomé, S. Fritz, and A. Belward, 2004. A new land-cover map of Africa for the year 2000, Journal of Biogeography, Vol.31 (6), 861–877.

Mbao, O.E., N. Kitaka, S.O Oduor, and J. Kipkemboi, 2012. Effects of nutrients on Periphyton community structure in Nyangores tributary, Mara river basin, Kenya. 3rd Lake Victoria Basin Scientific conference, Entebbe, Uganda.

Mburu, M.W.K., F.K. Lenga, and D.M. Mburu, 2011. Assessment of Maize yield response to Nitrogen fertilser in two semi-arid areas of Kenya with similar rainfall pattern, Journal of Agriculture, Science and Technology, 13 (1), 22-34.

Mbuvi, J.P., and E.B. Njeru, 1977. Soil resources of the Mau Narok area, Narok district; A preliminary investigation. Site evaluation report No.29, Kenya Soil Survey, S 410/KP/JBI-RFW.

McMaster, G.S., J.C. AscoughIi, D.A. Edmunds, A.A. Andales, L.E. Wagner, and F.A. Fox, 2005. Multi-crop plant growth modeling for agricultural models and decision support systems. In Proc. MODSIM 2005 Intl. Congress on Modelling and Simulation, 2138-2144.

Medlyn, B.E., C.V.M. Barton, M.S.J. Broadmeadow, R. Ceulemans, P. De Angelis, M. Forstreuter, M. Freeman, S.B. Jackson, S. Kellomaki, E. Laitat, A. Rey, P. Roberntz, B.D. Sigurdsson, J. Strassemeyer, K. Wang, P.S. Curtis, and P.G. Jarvis, 2001. Stomatal conductance of forest species after long-term exposure to elevated CO_2 concentrations: a synthesis, New Phytologist, 149, 247–264.

Mehmet C.D., A. Venancio, and E. Kahya, 2009. Flow forecast by SWAT model and ANN in Pracana basin, Portugal, Advances in Engineering Software, volume 40 (7).

Melesse, A.M., M. McClain, M. Abira, and W. Mutayoba, 2008. Hydrometeorological Analysis of the Mara River Basin, Kenya / Tanzania, World Environmental and Water Resources Congress, Ahupua'A.

Melone, F., S. Barbetta, T. Diomede, S. Peruccacci, M. Rossi, and A. Tessarolo, 2005. Review and selection of hydrological models – integration of hydrological models and meteorological inputs, RISK AWARE - INTEREG IIIB CADSES programme.

Migliaccio, K.W., and I. Chaubey, 2008. Spatial Distributions and Stochastic Parameter Influences on SWAT Flow and Sediment predictions, Journal of hydrologic engineering, 13 (4), 258-269.

Miller, S.N., W.G. Kepner, M.H. Mehaffey, M. Hernandez, R.C. Miller, D.C. Goodrich, K. K. Devonald, D.T. Heggem, and W.P. Miller, 2002. Integrating landscape assessment and hydrologic modeling for land cover change analysis, Journal of the American water resources association, Vol. 38 (4), 915-929.

Minaya, V., M.E. McClain, O. Moog, F. Omengo, and G. Singer, 2013. Scale-dependent effects of rural activities on benthic macro-invertebrates and physico-chemical characteristics in headwater streams of the Mara River, Kenya. Ecological indicators 32:116–122.

Mitchell, T., and T. Turner, 2006. Adapting to climate change; Challenges and opportunities for the development community, Tearfund.

MOA, 2011. Monthly market bulletin, Ministry of Agriculture, Kenya.

Montanari A., 2007. What do we mean by 'uncertainty'? The need for a consistent wording about uncertainty assessment in hydrology, Hydrological Processes, 21, 841–845.

Monteith, J.L., 1965. Evaporation and environment, Proceedings of the 19th Symposia of the Society for Experimental Biology,Cambridge, 205-234.

Moriasi, D.N., J.G. Arnold, M.W. Van Liew, R.L. Bingner, R.D. Harmel, and T.L. Veith, 2007. Model evaluation guidelines for systematic quantification of accuracy in watershed simulations. Transactions of the ASABE, Vol. 50 (3), 885–900.

Mote P. W., and G. Kasser, 2007. The shrinking glaciers of Kilimanjaro: can global warming be blamed? in F Gasse, Kilimanjaro's secrets revealed, Science, 298, 548–549.

Moulin, S., A. Bondeau, and R. Delecolle, 1998. Combining agricultural crop models and satellite observations: from field to regional scales.Int J Remote Sensing, 19 (6):1021-1036.

Muchela, A., 2012. Stochastic population forecast for Kenya: Arima model approach, UON.

Muchoney, D.M., and B.N. Haack, 1994. Change detection for monitoring forest defoliation. Photogrammetric Engineering and Remote Sensing 60:1243-1251.

Olsson, H., 1994. Changes in satellite-measured reflectances caused by thinning cuttings in boreal forests, Remote Sensing of Environment 50, 221-230.

Mulungu, D.M.M., and S.E. Munishi, 2007. Simiyu River catchment parameterization using SWAT model. Physics and Chemistry of the Earth, 32, 1032–1039.

Mundia C.N., and Y. Murayama, 2009. Analysis of Land Use/Cover Changes and Animal Population Dynamics in a Wildlife Sanctuary in East Africa, Remote Sensing, 1, 952-970.

Murphy, J., 1998. An Evaluation of statistical and dynamical techniques for downscaling local climate, Journal of climate, Volume 12, 2256-2283.

Mwiturubani D.A., 2010. Climate change and access to water resources in the Lake Victoria Basin, in: D.A. Mwiturubani, J.-A. van Wyk (Eds.), Climate change and natural resources conflicts in Africa, Vol. 170, Institute for Security Studies, Monograph, 63–77.

NASA, 2006. Landsat Program, Landsat ETM+, SLC-Off, USGS, Sioux Falls.

Nash, L.L., and P.H. Gleick, 1991. Sensitivity of streamflow in the Colorado Basin to climatic changes. J Hydrol 125:21

Nash, J.E., and J.V. Sutcliffe, 1970. River flow forecasting through conceptual models: Part 1. A discussion of principles. J. Hydrology 10(3), 282-290.

NBI, 2011. Kenya and the Nile basin initiative; Benefits of cooperation, NBI Secretariat.

Ndomba, P.M., and B.Z. Birhanu, 2008. Problems and prospects of SWAT model applications in Nilotic catchments: A Review, Nile Basin Water Engineering Scientific Magazine, Vol.1.

Ndomba, P.M., F.W. Mtalo, and A. Killingtveit, 2008. SWAT model application in a data scarce tropical complex catchment in Tanzania. Journal of Physics and Chemistry of the Earth, 33, 626-632.

Neitsch, S.L., J.G. Arnold, J.R. Kiniry, and J.R. Williams, 2007. Soil Water Assessment Tool– User Manual–Version 2005. Agriculture Research Service, Grassland, Soil and Water Research Laboratory,Temple,TX.

Neitsch, S.L., J.G. Arnold, J.R. Kiniry, R. Srinivasan, and J.R Williams, 2002. Soil and water assessment tool user's manual Version 2000. GSWRL Rep. No.02-02, BRCRep.No.02-06, TR-192,Texas Water Resources Institute, College Station,Tex.

NEMA, 2009. Bomet district environment action plan 2009-2013.

Nemani, R.R., C.D. Keeling, H. Hashimoto, M. Jolly, S.C. Piper, and C.J. Tucker, 2003. Climate driven increases in terrestrial net primary production from 1982 to 1999. Science, 300, 1560–1563.

Nepad, 2007. Climate Change and Africa, 8th Meeting of the Africa Partnership Forum Berlin, Germany

New, M., D. Lister, M. Hulme, and I. Makin, 2002. A high-resolution data set of surface climate over global land areas, Climate Res., 21, 1–25.

Nhemachena, C., and R. Hassan, 2007. Micro-level analysis of farmers' adaptation to climate change in Southern Africa, IFPRI Discussion Paper No. 714.

Nikolakopoulos, K.G., E.K. Kamaratakis, and N. Chrysoulakis, 2006. SRTM vs ASTER elevation products. Comparison for two regions in Crete, Greece, International Journal of Remote Sensing, Vol.27, (21), 4819–4838.

Notter, B., 2010. Water-related ecosystem services and options for their sustainable use in the Pangani Basin, East Africa, Switzerland, Institute of Geography, University of Bern, PhD thesis, 223 pp.

Notter, B., K. Abbaspour, H. Hurni, 2009. Setting up SWAT to quantify water-related ecosystem services in a large data-scarce watershed in East Africa. Proceedings of the 5th International SWAT Conference, Boulder, Colorado.

Nyong A. 2005. Abstract in Impacts of Climate Change in the Tropics: The African Experience. Avoiding Dangerous Climate Change, Meterological Office, Exeter, UK.

Nyoro, J. K., 2002. Kenya's competitiveness in domestic maize production: Implications for food security, Tegemeo Institute,Egerton University, Kenya.

Nzuma, J.M., and R. Sarker, 2010. An error corrected almost ideal demand system for major cereals in Kenya Agricultural Economics, 41, 43–50.

Nzuma, J. M., M. Waithaka, R.M. Mulwa, M. Kyotalimye, and G. Nelson, 2010. Strategies for adapting to climate change in rural Sub-Saharan Africa: A Review of data sources, Poverty Reduction Strategy Programs (PRSPs) and National Adaptation Plans for Agriculture (NAPAs) in ASARECA Member Countries, IFPRI Discussion Paper 01013.

Obare, L., and J.B. Wangwe, 1998. Underlying causes of deforestation and forest degradation in Kenya. NGO/IPO-led Underlying Causes Initiative regional workshop, Accra, Ghana, http://www.wrm.org.uy/deforestation/Africa.html

Onyando, J., D. Agol, and L. Onyango, 2013. WWF Mara River Basin Management Initiative, Kenya and Tanzania :Phase III – Final evaluation report. WWF_Kenya.

Osima, S., B. Hewitson, M. Stendel, 2012. Assessment of climate variability and change in East Africa in using high resolution models, HIRHAM5, 10th ICSHMO conference, Noumea, New Caledonia, France.

Park M., G. Park, S. Kim, 2010. Comparison of watershed sreamflow by using the projected MIROC3.2hires GCM data and the observed weather data for the period of 2000-2009 under SWAT simulation, 2010 international SWAT conference and Workshops.

Parry, M.L., O.F. Canziani, J.P. Palutikof and Co-authors: Technical Summary. Climate Change 2007: Impacts, Adaptation and Vulnerability. Contribution of Working Group II to the Fourth Assessment Report of the Intergovernmental Panel on Climate Change, M.L. Parry, O.F. Canziani, J.P. Palutikof, P.J. van der Linden and C.E. Hanson, Eds., Cambridge University Press, Cambridge, UK, 23-78. 2007

Penman, H.L., 1956. Estimating evaporation. Trans. American Geophysical Union, 37: 43-50.

Peters, A.J., E.A. Walter-Shea, L. JI, A. Vliia, M. Hayes, and M. D. Svoboda, 2002. Drought monitoring with NDVI-Based standardized vegetation index, Photogrammetric Engineering & Remote Sensing, Vol. 68 (1), 71-75

Pettorelli, N., J.O. Vik, A. Mysterud, G. Jean-Michel, C.J. Tucker, and N. C. Stenseth, 2005. Using the satellite-derived NDVI to assess ecological responses to environmental change and Evolution, TRENDS in Ecology, Vol.20 (9), 503-510.

Pineda, N., O. Jorba, J. Jorge, and J.M. Baldasano, 2004. Using NOAA AVHRR and SPOT VGT data to estimate surface parameters: Application to a mesoscale meteorological model, Int. J. Remote Sensing, 25 (1), 129–143.

Piper, A.M., 1944. A graphic procedure in the geochemical interpretation of water analyses, Trans.Am. Geophy. Union, 25, 914-928.

Pokorný, J., 2010. Direct role of plants, water,land in local climate. Towards sustainable use of ecosystems and agriculture. Canberra, CSIRO.

Praskievicz, S., and H. Chang, 2009. A review of hydrological modelling of basin-scale climate change and urban development impacts, Progress in Physical Geography, 33, 650–671.

Prathumratana, L., S. Sthiannopkao, K. W. Kim, 2008. The relationship of climatic and hydrological parameters to surface water quality in the lower Mekong River, Environment International, Volume 34 (6), 860–866.

Priestly, C.H.B., and R.J. Taylor, 1972. On the assessment of Surface Heat Flux and Evaporation Using Large-Scale Parameters, Monthly Weather Review Vol. 100 (2).

Pritchard, S.G., H.H. Rogers, S.A. Prior, and C.M. Peterson, 1999. Elevated CO_2 and plant structure: a review, Global Change Biology 5, 807–837.

Qi, C., and S. Grunwald, 2005. GIS-based hydrologic modeling in the Sandusky Watershed. Transactions of the ASAE, 48 (1), 169-180.

Qian, B., H. Hayhoe, and S. Gameda, 2005. Evaluation of the stochastic weather generators LARS-WG and AAFC-WG for climate change impact studies Clim Res, Vol. 29, 3–21.

Qian, B., S. Gameda, H. Hayhoe, R. De Jong, and A. Bootsma, 2004. Comparison of LARS-WG and AAFC-WG stochastic weather generators for diverse Canadian climates, Vol. 26, 175–191.

Rahbeh, M., D. Chanasyk, and J. Miller, 2011. Two-way calibration-validation of SWAT model for a small prairie watershed with short observed record, Canadian Water Resources Journal. 36 (3), 247-270.

Rajagopalan, B., U. Lall, and S. E. Zebiak, 2002. Categorical climate forecasts through regularization and optimal combination of multiple GCM ensembles. Monthly Weather. Review, 130, 1792–1811.

Ramakrishna, R. N., C.D. Keeling, H. Hashimoto, W.M. Jolly, S.C. Piper,C.J. Tucker, R.B. Myneni, and S.W. Running, 2003. Climate-driven increases in global terrestrial net primary production from 1982 to 1999, Science 300, 1560-1563.

Richards, J.F., 1990. Land transformation. In Turner 11, B.L., Clark, W.C., Kates, R.W., Richards, J.F., Mathews, J.T. and Meyer, W.B. (Eds): The Earth as Trans formed by Human Action. Cambridge University Press, Victoria, 161-178

Richard, Y., and I. Poccard, 1998. A statistical study of NDVI sensitivityto seasonal and interannual rainfall variations in southern Africa, Int. J.Remote Sens., 19, 2907–2920.

Richardson, C.W. and D.A. Wright, 1984. WGEN: a model for generating daily weather variables. US Department of Agriculture, Agricultural Research Service, ARS-8, 83 pp.

Riggan, Jr N.D., and R.C. Weih,Jr, 2009. A comparison of pixel-based versus object-based land use/land cover classification methodologies, Journal of the Arkansas Academy of Science Vol. 63, 145-152.

Riggs, H.C., 1972. Low flow investigations. Techniques of water resources investigations of the USGS, Book 4, Hydrological Analysis and interpretation, Washington DC, 18 pp.

Ringius, L., T.E. Downing, M. Hulme, D. Waughray and R. Selrod, 1996. Climate Change in Africa - Issues and Challenges in Agriculture and Water for Sustainable Development. Report 1996:8. Centre for International Climate and Environmental Research, Oslo. ISSN: 0804-4562.

Richter, B.D., J.V. Baumgartner, J. Powell, and D.P Braun, 1996, A method for assessing hydrologic alteration within ecosystems: Conservation Biology, Vol. 10 (1), 163–174.

Richter, B.D., J.V. Baumgartner, R. Wigington, and D.P. Braun, 1997. How much water does a river need?: Freshwater Biology, Vol. 37, 231–249.

Riungu, E., A.O. Onkware, and J.B. Okeyo-Owuor, 2012. An assessment of effects of land use changes from forestry to agriculture in the Mau on the ecosystem functioning of Upper Mara River. 3rd Lake Victoria Basin Scientific conference, Entebbe, Uganda.

Roerink, G. J., M. Menenti, W. Soepboer, and Z. Su, 2003. Assessment of climate impact on vegetation dynamics by using remote sensing. Phys. Chem. Earth 28, 103-109.

Rogan, J., and D. Chen, 2004. Remote sensing technology for mapping and monitoring land-cover and land-use change, Progress in Planning, 61, 301–325.

Rouse, J.W., R.H. Haas, J.A. Schell, and D.W. Deering, 1973. Monitoring vegetation systems in the Great Plains with ERTS, Third ERTS Symposium, NASA SP-351 I, 309-317.

Royston, J.P., 1995. Remark AS R94: A remark on Algorithm AS 181: The W test for normality.Applied Statistics, 44, 547-551

Ryan, J., G. Estefan, and A. Rashid, 2001. Soil and plant analysis lab manual. 2nd ed. International Center for Agricultural Research in the Dryland Areas (ICARDA), Aleppo, Syria. National Agricultural Research Center, Islamabad, Pakistan.

Sader, S.A. and J.C. Winne, 1992. RGB-NDVI colour composites for visualizing forest change dynamics, International Journal of Remote Sensing, 13:3055-3067.

Sader, S.A., M. Bertrand, E.H. Wilson, 2003. Satellite change detection of forest harvest patterns on an Industrial forest landscape. Forest Science 49 (3), 341-353.

Sader, S A , D J Hayes, J.A. Hepinstall, M. Coan, and C. Soza, 2001. Forest change monitoring of a remote biosphere reserve, International Journal of Remote Sensing, 22 (10), 1937–1950.

Saggerson, E.P., 1991. Geology of the Nairobi area, Mines and geology department, Nairobi, Kenya, 89pp.

Saghravani S.R., S. Mustapha, S. Ibrahim, and E. Randjbaran, 2009. Comparison of daily and monthly results of three evapotranspiration models in tropical zone: A Case Study, American Journal of Environmental Sciences 5 (6), 698-705.

Salami, A., A.B. Kamara, and Z. Brixiova, 2010. Smallholder Agriculture in East Africa: Trends, Constraints and Opportunities, Working Papers Series N° 105 African Development Bank, Tunis, Tunisia.

Salmi, T., A. Maata, P. Antilla, T. Ruoho-Airola, and T. Amnell, 2002. Detecting trends of annual values of atmospheric pollutants by the Mann–Kendall test and Sen's slope estimates – the Excel template application Makesens. Finnish Meteorological Institute, Helsinki, Finland, 35 pp

Sannier, C.A.D., 1998. Real time vegetation monitoring with NOAA-AVHRR in southern Africa for wildlife management and food security assessment, International Journal of Remote Sensing, 19 (4), 621-639.

Savitzky, A., and M. J. E. Golay, 1964. Soothing and differentiation of data by simplified least squares procedures, Anal. Chem., Vol. 36, 1627–1639.

Schipmann, C., 2011. Variety Adoption of Orphan Crops by Smallholder Farmers in Tanzania, a Survey Based Choice Experiment Tropentag, Bonn.

Schwarz, G.E., A.B. Hoos, R.B. Alexander, and R.A. Smith, 2006. The SPARROW Surface Water-Quality Model: Theory, Application and User Documentation. U. S. Geological Survey Techniques and Methods Book 6, Section B, Chapter 3.

Scholes, B., A. Ayite-Lo, T. Nyong, R. Tabo, and C. Vogel, 2007. Global environmental change (including Climate Change and Adaptation) in sub-Saharan Africa, ICSU publications

Schulze, E.D., F.M. Kelliher, C. Körner, J. Lloyd, and R. Leuning, 1994. Relationships among maximum stomatal conductance, ecosystem surface conductance, carbon assimilation rate, and plant nitrogen nutrition: a global ecology scaling exercise. Annual Review of Ecology and Systematics 25: 629-660.

Schuman, H., and J. Converse, 1971. The effects of black and white interviewers on black responses. Public Opinion Quarterly, 35 (1).

Schuol, J., and K.C. Abbaspour, 2006. Calibration and uncertainty issues of a hydrological model (SWAT) applied to West Africa, Adv. Geosci. 9, 137–143.

Semenov, M. A., and E.M. Barrow, 2002. LARS-WG A Stochastic Weather Generator for Use in Climate Impact Studies Version 3.0 ,User Manual.

Semenov, M.A., R.J. Brooks, E.M. Barrow, and C.W. Richardson, 1998. Comparison of WGEN and LARS-WG stochastic weather generators for diverse climates, Climate Research, 10, 95-107.

Serneels, S., and E.F.Lambin, 2001. Proximate causes of land-use change in Narok District, Kenya: A spatial statistical model, Agriculture, EcosystemsandEnvironment, 85, 65–81.

Serneels, S., M.Y. Said, and E.F. Lambin, 2001. Land cover changes around a major east African wildlife reserve: The Mara Ecosystem (Kenya), International Journal of Remote Sensing, 22:17, 3397-3420.

Sevat, E., and A. Dezetter, 1991. Selection of calibration objective functions in the context of rainfall-runoff modeling in a Sudanese savannah area. Hydrological Sci. J. 36(4): 307-330.

Singh, P., S. Roy, and F. Kogan, 2003. Vegetation and temperature condition indices from NOAA AVHRR data for drought monitoring over India. Int. J. Remote sensing, Vol.24, NO.22, 4393–4402.

Sivapalan, M., K. Takeuchi, S.W. Franks, K. Gupta, H. Karambiri, V. lakshmi, X. Liang, J.J. McDonnell, E.M. Mendiondo, P.E. O'Connell, T. Oki, J.W.P. meroy, D. Schertzer, S. Uhlenbrook, and E. Zehe, 2003. IAHS Decade on Predictions in Ungauged Basins (PUB), 2003–2012:Shaping an exciting future for the hydrological sciences,Hydrol.Sci.J. 48 (6), 857-88.

Sonali, P., and D.N Kumar, 2013. Review of trend detection methodsand their application to detect temperature changes in India, J. Hydrol., 476, 212–227.

Sophocleous, M.A., J.K Koelliker, R.S Govindaraju, T.Birdie, S.R Ramireddygari, and .P. Perkins, 1999. Integrated numerical modeling for basin-wide water management: the case of the Rattlesnake Creek Basin in south-central Kansas. Journal of Hydrology 214 (1–4), 179–196.

Srinivasan, R., X. Zhang, and J. Arnold, 2010. SWAT ungauged: hydrological budget and crop yield predictions in the Upper Mississippi River Basin, Trans. ASABE, 53 (5): 1533–1546.

Storey, J., P. Scaramuzza, G. Schmidt, and J. Barsi, 2005. Landsat 7 scan line corrector-off gap filled product development, Pecora 16 "Global Priorities in Land Remote Sensing", Sioux Falls, South Dakota.

Swain, P.H., and S.M Davis, 1978. Remote Sensing: The quantitative approach, New York:McGraw-Hill.

Swallow, B.M., J.K. Sang, M. Nyabenge, D.K. Bundotich, A.K. Duraiappah, and T.B. Yatich, 2009. Tradeoffs, synergies and traps among ecosystem services in the Lake Victoria basin of East Africa. Environmental science and policy 1 2, 504– 519.

Tans. P., and R. Keeling, 2013. Trends in Atmospheric Carbon Dioxide.NOAA/ESRL (www.esrl.noaa.gov/gmd/ccgg/trends/) Scripps Institution of Oceanography (scrippsco2.ucsd.edu/).

Tebaldi C., and R. Knutt, 2007. The use of the multi-model ensemble in probabilistic climate projections, Phil.Trans.R.Soc.A , 365, 2053–2075.

Tetratech., 2004. White Paper Use of SWAT to Simulate Nutrient Loads and Concentrations in California, Tetra tech progress report, http://rd.tetratech.com/epa/.

Tisseuil C., M. Vrac, S.Lek, and A.J. Wade, 2010. Statistical downscaling of river flows, Journal of Hydrology, 385, 279–291.

Toukiloglou P., 2007. Comparison of AVHRR, MODIS and VEGETATION for land cover mapping and drought monitoring at 1 km spatial resolution School of Applied Sciences, Cranfield University, PhD Thesis

Tou, J.T., and R.C. Gonzales, 1974. Pattern recognition principles. Iso-data algorithm, Pattern Classification by Distance Functions, Addison-Wesley, Reading, Massachusetts, 97-104.

Tulu D.M., 2005. SRTM DEM Suitability in Runoff Studies, ITC, the Netherlands, Msc Thesis.

Turner B.L., D. Skole, S. Sanderson, G. Fischer, L. Fresco, and R. Leemans, 1995. Land-Use and Land-Cover Change Science/Research Plan. IHDP Report No. 07, http://www.ihdp.uni-onn.de/html/publications/reports/report07/luccsp.htm.

Uchinda S., 2001. Discrimination of agricultural land use using Multi-temporal NDVI data. 22nd Asian conference on remote sensing, Singapore.

UNFCCC., 2007. Baseline Socio-economic Scenarios, UNFCCC.

Urama K.C., G. Davidson, and S. Langan, 2008. Towards an Integrated Trans-boundary River Management Policy Development in SemiArid River Basins, Proceedings of a Pan- Proceedings of a Pan-African Stakeholder Policy Forum, Arusha, Tanzania

Urama K. C., and G. Davidson, 2008. Bordering on a Water Crisis: The Need for Integrated Resource Management in the Mara River Basin, A Policy Brief.

USACE., 1995. Hydrologic Engineering Center's Hydrologic Modeling System (HEC-1/HEC-HMS, , US Army corps of Engineers, technical paper no. 150.

USGS., 2012. Landsat: A Global Land-Imaging Mission. http://landsat.usgs.gov.

Van Griensven A., F. Kilonzo, P. Ndomba, and P. Gassman, 2011. Current application trends and data challenges regarding the use of the Soil and Water Assessment Tool (SWAT) in Africa, International SWAT Conference, Toledo, Spain.

Van Griensven A., M.K. Akhtar, D. Haguma, R. Sintayehu, J. Schuol, K.C. Abbaspour, S.J. Van Andel, and R.K. Price, 2007. Catchment modeling using internet based global data, 4th SWAT international conference.

Van Griensven A., T. Meixner, S. Grunwald, T. Bishop, M. Diluzio, and R. Srinivasan, 2006. A global sensitivity analysis tool for the parameters of multi-variable catchment models, Journal of Hydrology, 324, 10–23.

Van Griensven, A., and T. Meixner, 2006. Methods to quantify and identify the sources of uncertainty for river basin water quality models, Water Science and Technology 53 (1), 51–59.

Van Liew, M.W., T.L Veith, 2009. Guidelines for Using the Sensitivity Analysis and Auto-calibration Tools for Multi-gage or Multi-step Calibration in SWAT.

Vazquez-Amábile G.G., and B.A. Engel, 2005. Use of SWAT to compute groundwater table depth and streamflow in the Muscatatuck river watershed, Transactions of the ASAE, Vol. 48 (3), 991–1003.

Verburg, P.H., and A. Veldkamp, 2004. Projecting land use transitions at forest fringes in the Philippines at two spatial scales. Landscape Ecology,19 (1), 77–98.

Verburg, P.H., P.P. Schot., M.J. Dijst., and A. Veldkamp, 2003. Land use change modeling: current practice and research priorities. CABI series.

Verhoef, W., 1996. Application of Harmonic Analysis of NDVI Time Series (HANTS), in Fourier Analysis of Temporal NDVI in the Southern African and American Continents, S.A.a.M. Menenti, Edsr: DLO Winand Staring Centre, Wageningen, TheNetherlands, 19-24.

von Stackelberg, N.O., G.M. Chescheir, R.W. Skaggs, and D.M. Amatya, 2007. Simulation of the hydrologic effects of afforestation in the Tacuarembo River basin, Uruguay, Transactions of the ASABE, Vol. 50 (2), 455–468.

Vörösmarty C.J., P.G.J. Salisbury, and R.B. Lammers, 2000. Global Water Resources: Vulnerability from Climate Change and Population Growth Science, 289-284.

Walsh R.P.D, D.M Lawler, 1981. Rainfall Seasonality Description, Spatial patterns and change through time. Weather, 36: 201-208.

Wang, J., T.W. Sammis, V.P. Gutschick, M. Gebremichael, S.O. Dennis, and R.E. Harrison, 2010. Review of Satellite Remote Sensing Use in Forest Health Studies, The Open Geography Journal, 3, 28-42.

Wangai, A.W., M.G. Redinbaugh, Z.M. Kinyua, D.W. Miano, P.K. Leley, and M. Kasina, G. Mahuku, K. Scheets, and D. Jeffers, 2012. First Report of Maize chlorotic mottle virus and Maize Lethal Necrosis in Kenya, Plant disease Volume 96 (10), 1582-1582.

Washington, R., M. Harrison, D. Conway, E. Black, A. Challinor, D. Grimes, R. Jones, A. Morse, G. Kay, and M. Todd, 2006. African climate change:Taking the Shorter Route, American meteorological society, 1355-1366.

Watson, B.M., N. Coops, S. Selvalingam, and M. Ghafouri, 2005. Integration of 3-PG into SWAT to simulate growth of evergreen forests in Australia, 3rd International SWAT Conference, Zurich, Switzerland.

Watson, B.M., S. Selvadore, and M. Ghafouri, 2003. Evaluation of SWAT for modelling the water balance of the Woady Yaloak River catchment, Victoria, in MODSIM 2003: International Congress on Modelling and Simulation, Jupiters Hotel and Casino, 14-17 July 2003, A.C.T, 1-6.

Wattenbach, M., F. Hatterman, R. Weng, F. Wechsung, V. Krysanova, and F. Badeck, 2005. A simplified approach to implement forest eco-hydrological properties in regional hydrological modeling. Ecol. Model. 187 (1), 4950.

White, E.D., Z.M. Easton, D.R. Fuka, A.S. Collick, E. Adgo, M. McCartney, S.B. Awulachew, Y.G. Selassie, and T.S. Steenhuis, 2011. Development and application of a physically based landscape water balance in the SWAT model, Hydrol. Process, 25, 915–925, doi:10.1002/hyp.7876,

White, M.J., R.D., Harmel, J.G. Arnold, and J.R. Williams, 2012. SWAT Check: A screening tool to assist users in the identification of potential model application problems, Journal of Environmental Quality.

WHO., 2008. Guidelines for drinking-water quality [electronic resource]:incorporating 1st and 2nd addenda,Vol.1,Recommendations.– 3rd ed. NLM classification: WA 675 Geneva

Wilby, L., 2005. Uncertainty in water resource model parameters used for climate change impact assessment Robert, Hydrological Processes, 19, (16), 3201–3219.

Wilby, R.L., L.E Hay., W.J. Gutowski Jr, R.W. Arritt, E.S Takle, Z. Pan, G.H Leavesley, and M.P. Clark, 2000. Hydrological responses to dynamically and statistically downscaled Climate model output, Geophysical Research Letters, 27, 1199–1202.

Wilby, R.L., and T.M.L. Wigley, 1997. Downscaling general circulation model output: A review of methods and limitations, Prog. Phys. Geogr., 21, 530-548.

Wilks, D.S. and R.L. Wilby, 1999. The weather generation game: a review of stochastic weather models.Progress in Physical Geography 23: 329-357.

Willems P., 2009. A time series tool to support the multi criteria performance evaluation of rainfall-runoff models, Environmental Modelling & Software 24: 311–321.

Williams, J.R., and R.W. Hann, 1972. Hymo, A problem-oriented computer language for building hydrologic models, Water Resources Research, Volume 79–86.

Williams, J.R., C.A. Jones, J.R. Kiniry, and D.A. Spanel, 1989. The EPIC crop growth model. Trans. ASAE 32, 97–511.

Wilschut, L.I., 2010. Land use in the Upper Tana. Technical report of a remote sensing based land use map. Green Water Credits Report 9 / ISRIC Report 2010/03.ISRIC–World Soil Information, Wageningen.

Wilson, K.B., T.N. Carlson, and J.A. Bunce, 1999. Feedback significantly influences the simulated effect of CO_2 on seasonal evapotranspiration from two agricultural species.Global Change Biology, 5, 903–917.

Witt, C., J.M. Pasuquin, and G. Sulewski, 2009. Predicting agronomic boundaries of future fertilizer needs in AgriStats. Better Crops, Vol. 93, No. 4, 16-18.

Wood, J.D., 1996. The geomorphological characterisation of digital elevation models, University of Leicester, UK, PhD Thesis.

Wood, A.W., L.R. Leung, V. Sridhar, and D.P. Lettenmaier, 2004. Hydrologic implications of dynamical and statistical approaches to downscaling climate model outputs, Climate Change, 62 (1-3), 189-216.

Wood,A.W., E.P. Maurer, A. Kumar, ,and D.P. Lettenmaier, 2002. Long range experimental Hydrologic forecasting for the Eastern U.S., Geophysical Research Letters 107 (20),4429.

Wortmann, C., M. Mamo, C. Mburu, E. Letayo, G. Abebe, K.C. Kayuki, M. Chisi, M Mativavarira, S. Xerinda, and T. Ndacyayisenga, 2009. Atlas of Sorghum Production in Eastern and Southern Africa. The Board of Regents of the University of Nebraska, University of Nebraska, Lincoln.

Wullschleger, S.D., C.A. Gunderson, P.J. Hanson, K.B. Wilson, and R.J. Norby, 2002. Sensitivity of stomatal and canopy conductance to elevated CO_2 concentration interacting variables and perspectives of scale, New Phytologist, 153, 319–331.

Xu, C., 1999. From GCMs to river flow: a review of downscaling methods and hydrologic modelling approaches, Progress in Physical Geography, 23: 229.

Yang, J., P. Reichert, K.C. Abbaspour, J. Xia, and H. Yang, 2008. Comparing uncertainty analysis techniques for a SWAT application to the Chaohe Basin in China, Journal of Hydrology, 358, 1–23.

Yang, K., P. Reichert, K.C. Abbaspour, H. Yang, 2007. Hydrological modelling of the Chaohe Basin in China: statistical model formulation and Bayesian inference, Journal of Hydrology, 340, 167–182.

Yates, D., 1996. Integrating Water into an Economic Assessment of Climate Change Impacts on Egypt, IIASA Working Paper WP-96-031.

Yue S., and C. Wang, 2004. The Mann-Kendall Test Modified by Effective Sample Size to Detect Trend in Serially Correlated Hydrological Series. Water Resources Management, 18, 201–218.

Yue, S., P. Pilon, and G. Caradias, 2002. Power of theMann-Kendall and Spearman's rho tests for detecting monotonic trends in hydrological series, J. Hydrol., 259, 254-271.

Yin, H., T. Udelhoven, R. Fensholt, D. Pflugmacher, and P. Hostert, 2012. How Normalized Difference Vegetation Index (NDVI) Trends from Advanced Very High Resolution Radiometer (AVHRR) and Système Probatoire d'Observation de la Terre

VEGETATION (SPOT VGT) Time Series Differ in Agricultural Areas: An Inner Mongolian Case Study, Remote Sens., 4, 3364-3389.

Young, R.A., C.A. Onstad, D.D. Bosch, and W.P. Anderson, 1987. AGNPS, Agricultural Nonpoint-Source Pollution Model: A Watershed Analysis Tool, U.S. Department of Agriculture, Washington, DC.

Zhang X.S., R. Srinivasan, B. Debele, F.H Hao, 2008. Runoff simulation of the headwaters of the Yellow Riverusing the SWAT model with three snowmelt algorithms, J Am Water Resour Assoc. 44 (1), 48–61.

Zheng, G., and L.M. Moskal, 2009. Retrieving Leaf Area Index (LAI) Using Remote Sensing: Theories, Methods and Sensors, Sensors, 9, 2719-2745.

Zhou, G., N. Minakawa, A.K. Githeko, and G. Yan, 2004. Association between climate variability and malaria epidemics in the East African highlands, Proceedings of the National Academy of Sciences of the United States of America. 101: 2375-2380.

Zhou, L., R. K Kaufmann, Y. Tian, R. B. Myneni, and C. J Tucker, 2003. Relation between interannual variations in satellite measures of northern forest greenness and climate between 1982 and 1999, Journal of Geophysical Research, 108.

Zou H., and Y. Yang, 2004. Combining time series models for forecasting, Interrnational Journal of Forecasting, 20, 69–84

Zsembeli, J., G. Kovács, and A. Mándoki, 2011. Water use efficiency of maize and different sorghum hybrids under lysimeter conditions, Gumpensteiner Lysimetertagung, 227–230.

14. Annexes

Annex I: Open source software/tools used in the study

Software/ tool	Functions used	Brief description
1 Timesat	1. Time series extraction 2. Filter time series	The Timesat 3.0 software manual consists of three parts. Part I gives general information together with examples of some applications of Timesat. Part II describes the algorithms underlying the software package. Also the settings affecting the processing are discussed in detail. Part III is the software user's guide, with detailed information on how to install, run, and handle the program package. The Timesat program package is designed primarily for analyzing time-series of satellite data and uses an adaptive Savitzky-Golay filtering method and methods based on upper envelope weighted fits to asymmetric Gaussian and double logistic model functionsFrom the fitted model functions a number of seasonality parameters, e.g. beginning and end of the growing season, can be extracted. Parameters for a number of pixels can be merged into a map displaying seasonality on a regional or global scale. Timesat consists of a number of numerical and graphical routines coded in Matlab & Fortran 2003. The Matlab version of the package, which comes with a versatile graphical user interface (GUI), is mainly applied for testing input settings and running smaller cases. Matlab routines are pre-compiled and users of the program package do not need to have Matlab installed. Instead processing is done through a runtime engine called the Matlab Compiler Runtime (MCR), that is set up on the users machine by executing the file MCRinstall.exe The latter file is provided along with the Timesat program package. The Fortran

version of the package is highly vectorized and efficient and should used to process large data sets. Also for the Fotran version we supply pre-compiled executable files.

| 2 | VGTExtract | Extract SPOT-VGT files | The primary aim of VGTExtract is to facilitate the integration of SPOT-VGT products into commonly used GIS and Remote Sensing software for further visualisation, analysis and postprocessing. VGTExtract searches for VGT products in a given directory and its subdirectories. For each product found, VGTExtract can automatically unpack (uncompress) the product and perform the following actions on the resulting unzipped HDF data files:

Mosaic the data layers, when they are provided as a set of regular tiles (e.g. ten by ten degrees, as in Geoland2 products);

Extract a given rectangular geographic bounding box or Region of Interest (ROI);

Convert to a set of file formats;

Change data type as may be required by the output format;

Apply a linear encoding of the form output = (scale * input) + offset to all values except one (and only one) special value for missing data.

When extracting the geographic region from an input image, VGTExtract checks if the desired output region intersects with the region covered by the input image.

When this intersect is only partial, VGTExtract automatically fills the non-intersecting part of the output image with missing data values.

This means that the output file will always have the size of the desired region, but it may be partly empty in those areas that were not present in the input image. |
| 3 | 3DEM | Convert HDF to GEOTIFF | The program that opens DEM data and allows visualization and use it in a number of different ways. It comes with a first-rate PDF manual, available from the Help menu. 3DEM will accepts DEM formats, like USGS DEMs (ASCII and SDTS), .hgt, .bil, LIDAR .las, even MARS MOLA (.img).

It allows for selection and download of lower-resolution DEM data from the GLOBE merge them together, and allows saving of merged form in either USGS ASCII |

format or GeoTiff format.

4	VTbuilder	Render Eind Bytes	VTBuilder is a tool for viewing and processing geospatial data. It can import a wide variety of data formats, and output efficiently to 3d runtime software. The usual process of vizualising DEMs 1. Acquire raw geospatial data (elevation, road vectors, etc.) 2. Read them into VTBuilder. 3. Clean up the data with operations such as bringing it into alignment, extracting areas of interest, merging and resampling, and supplying missing information. 4. Write the data out
5	PAST	Autocorrelation Durbin Watson statistics	The PAST program integrates spreadsheet-type data entry with univariate and multivariate statistics, curve fitting, time-series analysis, data plotting, and simple phylogenetic analysis
6	Makesen	Mann-Kendall trends analysis Sens estimator	An Excel template – MAKESENS – is developed for detecting and estimating trends in the time series of annual values of atmospheric and precipitation concentrations. The procedures based on the nonparametric Mann-Kendall test for the trend and the nonparametric Sen's method for the magnitude of the trend. The Mann-Kendall test is applicable to the detection of a monotonic trend of a time series with no seasonal or other cycle. The Sen's method uses a linear model for the trend

Annex 2: Kensoter soil classes in the study area

	NEWSUID	REVISED FAO1988 CODE	TEXTURAL CLASS	KSS CODE	%
1	45	humic Acrisol	sandy loam	F17	2.59
2	56	eutric Regosol	silt loam	H13	1.29
3	57	Cambisol	clay	L10	1.07
4	58	humic Regosol	clay	L25	2.3
5	61	humic Phaeozem	clay	A14	1.95
6	93	humic Phaeozem	sandy clay loam	F4	5.49
7	183	Nitisol	clay	Ps7	0.66
8	187	vertic Luvisol	clay	Up2	9.93
9	190	Nitisol	clay	Ps7	0.94
10	192	mollic Andosol	clay loam	Pc6	9.32
11	196	mollic Andosol	loam	F10	35.47
12	200	mollic Andosol	clay	Uc8	1.16
13	377	eutric Planosol	clay loam	Pv11	0.31
14	378	humic Phaeozem	sandy clay loam	L20	0.92
15	380	eutric Vertisol	clay	Pn4	5.31
16	381	eutric Planosol	clay loam	Up5	1.19
17	382	humic Regosol	loam	Pn11	4.13
18	386	Cambisol	loam	H16	4.29
19	387	eutric Planosol	clay loam	Pn10	0.06
20	389	eutric Planosol	clay loam	Pn6	5.2
21	391	eutric Planosol	clay loam	Pn7	6.41

Annex 3a: 1978 Survey of Kenya topographic sheets used in the accuracy assessment of 1976/86 thematic maps

262

Annex 3b: Ground control points (GCPs) used for the ground truthing of the NDVI map and 2006 thematic map

Annex 4: Sample questionaire for the Upper Mara study

INFORMED CONSENT   **Code:**

FK/_____/

I have been asked to participate in a questionnaire activity that forms part of the PhD studies for **Fidelis N. Kilonzo** of **Kenyatta University**, P.O BOX 43844-00100, Nairobi. This questionnaire has been designed to gather information about the **Crop yield and related farm management practices within Mara river basin**. I understand that any information I provide shall be kept confidential, and the results that may be published in a professional or academic report or journals will be anonymous.

I hereby consent to participate in this survey under the specified conditions.

Name:_____

Signature (Thumbprint) _____ Date

_____

Interviewer code: FK/INTW
Nr./_____

1. Farm details1.1Where is the farm located?

Description	Name
Village	
Sub-location	
Location	
Division	
District	
County	
Geog. coordinates	

1.2 Who is responsible for the farm? |__|

1=	Husband	4=	Relative
2=	Wife	5=	Other(specify)
3=	Child		

2.1 What type of farming system do you practice? |__|

Type of farming system	
1= mixed farming 3=	Intercropping
2= Monoculture 4=	Other (specify)

2.3.What are the approximate sizes of the of the parcel/block?
Unit for area: _____

Land piece (Name/location)	Size
1	
2	
2	

3. Types of crops grown

3.1. Did the household plant any crops during the last rainy season? |__|
(Yes = 1, No = 2), *if no go to 6*

3.2. For each crop planted during last rainy season provide the following information:

	Crop name		Planting		Harvesting	
	Local	**English**	**Dates**	**Area**	**Dates**	**Quantity**
1.						
2.						
3.						
4.						
5.						

Units for area _____ Conversion   to   S.I

Units   for   harvest   _____ Conversion   to   S.I

4. Management practices4.1. How do you improve the fertility of your farm?
Fertilizer |__| Manure |__| Both Fertilizer+ Manure |__| None |__|
If fertlizer go to 4.1.1,
if manure go to 4.1.24.1.1.

List here the type and quantity of fertilizers used

Fertilizer type	Quantity used

4.1.2. What is the source and quantity of manure used ?4.1.2.1. What is the source of manure used ?
Composting |__| Farmyard |__| Animal dung |__| Other specify |__|
4.1.2.2. What is the quantity of manure used ? _____

4.2. How do you control weeds in your farm?
Weeding |__| Herbicides |__| Weeding +Herbicides |__| None |__|
If herbicides, go to 4.2.1
4.2.1 List here the type and quantity of herbicides used

Herbicides type	Crop used on	Quantity used

4.3. How do you control pests in your farm?

Pesticides |__| organic |__| None |__|

If pesticides, go to 4.3.1.

If organic , go to 4.3.2

4.3.1 List here the type and quantity of Pesticides used

Pesticides type	Crop used on	Quantity used

4.3.2 List here the type of organic method used

Organic method	Crop used on	Quantity used

4.4. What do you do with the crop residuals after harvesting the grain?

Crop residual			
1=	left to rot insitu	3=	burnt
2=	fed to own animals	4=	Other (specify)

5. Market for the produce; Licensed marketer/wholesale |__| Broker |__|
Retail |__|
6. What is reasons for not growing crops and vegetables during the last rainy season?
|__|

Main reasons for not growing crops and vegetables during last rains					
01=	no rains	03=	no inputs	05=	Other (specify)
02=	rains came rate	04=	illness/social problems		

7. Type of tree crops
7.1. Does the farm grow any tree crops? 1 = Yes, 2 = No |__|
If yes, go to 7.2

7.2. What type of tree crops does the farm grow and what number of trees?

S/NO	Tree type	Number of trees	where planted	main use	other use	Amount harvested	harvest period
1							
2							
3							
4							
5							

		permanent crops		Forest trees and plantations	
Fruits and nuts					
100 =	Mango	200=	Tea	300=	Natural
110 =	Oranges	210=	Coffee	310=	Plantation
120 =	Passion	220=	Other (specify)	320=	Other (specify)
130 =	Pawpaw				
140 =	Other (specify)				

where planted
1=   mostly on field/plot boundaries
2=   mostly scattered in fields
3=   mostly in plantation

Uses
1=   timber   4= shade
2=   food   5= medicinal
3=   charcoal   6= Other (specify)

8. Animals kept in the farm

8.1 Does the farm have any domestic animals? 1 = Yes, 2 = No |__|

If yes, go to 8.2

8.2 Which animals are kept and their number?

Type	Number	Zero grazing?	Type	Number	Zero grazing ?
Cattle			Chicken		
Pigs			Rabbit		
Goats			Donkey		
Sheep			Ducks		

9.0 Any other observations/Remarks from farmer

_____
_____
_____
_____

10.0 Any special notes by the Interviewer

_____

Annex 5. Description of all the sites sampled in the study, including the main 16 sites labeled K1-K18

ID	Site	Name	longitude	Latitude	Subbasin	Landuse
Main sites*						
1	K1	Emarti	35.23E	1.06S	Mara	Agriculture
2	K2	Lelaitich	35.64E	0.7S	Amala	Agriculture
3	K3	Kimulgul	35.55E	0.79S	Amala	Forest
4	K6	Nyangores	35.42E	0.7S	Nyangores	Forest
5	K7	Lionget	35.5E	0.58S	Nyangores	Agriculture
6	K8*	Masese	35.44E	0.73S	Nyangores	Forest
7	K9*	Matecha	35.5E	0.83S	Amala	Forest
8	K10*	Bomet	35.35E	0.79S	Nyangores	Agriculture
9	K11	Kaptwek	35.66E	0.48S	Nyangores	Agriculture
10	K12	Olposmoru	35.78E	0.55S	Amala	Agriculture
11	K13*	Mulot	35.42E	0.94S	Amala	Agriculture
12	K14*	Silibwet	35.36E	0.74S	Nyangores	Agriculture
13	K15	Oldmara	35.04E	1.22S	Mara	Mixed
14	K16*	Kapkimolwa	35.44E	0.9S	Amala	Mixed
15	K17	Baraget	35.8E	0.45S	Nyangores	Forest
16	K18	Tipis	35.71E	0.56S	Amala	Mixed
Other sites**						
17		Ainabcheruik	38.48E	0.68S	Nyangores	Forest
18		Ainabngetuny	35.44E	0.72S	Nyangores	Forest
19		Borowet	35.42E	0.71S	Nyangores	Agriculture
20		Chepkosiom I	35.45E	0.7S	Amala	Agriculture
21		Chepkosiom II	35.42E	0.71S	Amala	Agriculture
22		Chepkosiom III	35.42E	0.71S	Amala	Agriculture

*the 6 Key sites used for physical-chemical analysis

** for nutrient monitoring

Annex 6a: Script used in ERDAS model maker for calculation of VCI

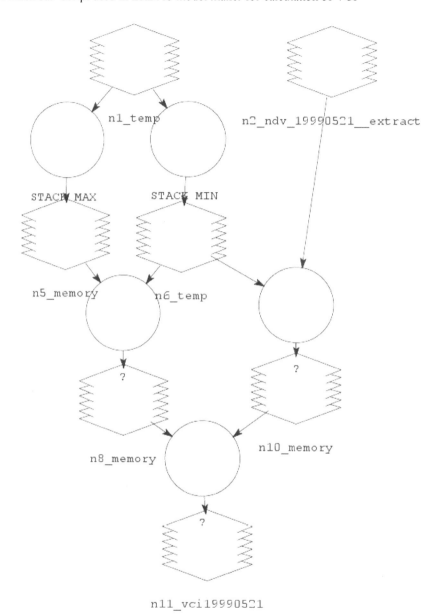

Annex 6b: Script used in ERDAS model maker for calculation of Z scores in SVI determinations

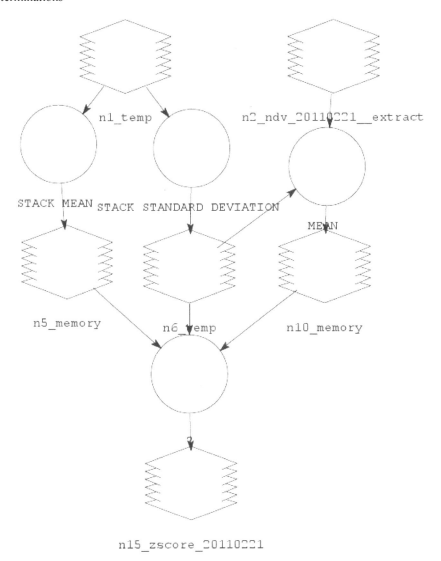

n1_temp

n2_ndv_20110221__extract

STACK MEAN   STACK STANDARD DEVIATION

MEAN

n5_memory   n6_temp   n10_memory

n15_zscore_20110221

Annex 7. The spatial reference properties

Feature	Description
Name	WGS_1984_Mercator
Projection	Mercator
False_Easting	0.000000
False_Northing	0.000000
Central_Meridian	0.000000
Standard_Parallel_1	0.000000
Linear Unit	Meter (1.000000)
Geographic Coordinate System	GCS_WGS_1984
Angular Unit	Degree (0.017453292519943299)
Prime Meridian	Greenwich(0.000000000000000000)
Datum	D_WGS_1984
Spheroid	WGS_1984
Semimajor Axis	6378137.00000000000000000

Annex 8: Management file edits for MAIZ and CORN HRUs

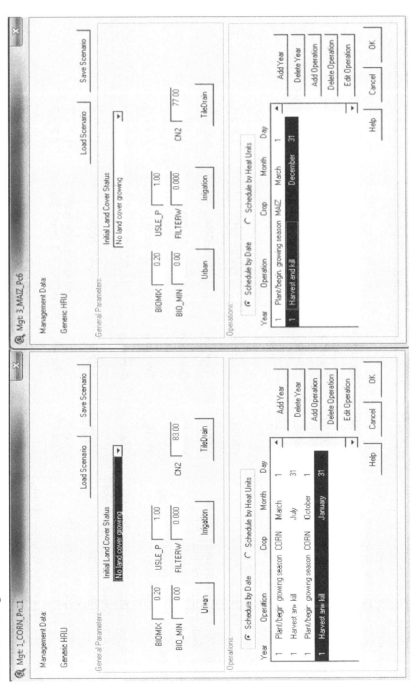

Annex 9. MMD dataset used in the LARS-WG

Centre	Centre acronym	Country	Global Climate Model	Grid resolution
Australia's Commonwealth Scientific and Industrial Research Organisation	CSIRO	Australia	CSIRO-MK3.0	1.9 x 1.9
Canadian Centre for Climate Modelling and Analysis	CCCma	Canada	CGCM3 (T47) CGCM3 (T63)	2.8 x 2.8 1.9 x 1.9
Beijing Climate Centre	BCC	China	BCC-CM1	1.9 x 1.9
Institute of Atmospheric Physics	LASG	China	FGOALS-g1.0	2.8 x 2.8
Centre National de Recherches Meteorologiques	CNRM	France	CNRM-CM3	1.9 x 1.9
Institute Pierre Simon Laplace	IPSL	France	IPSL-CM4	2.5 x 3.75
Max-Planck Institute for Meteorology	MPI-M	Germany	ECHAM5-OM	1.9 x 1.9
Meteorological Institute, University of Bonn	MIUB	Germany	ECHO-G	3.9 x 3.9
Model and Data Group at MPI-M	M&D	Germany	ECHO-G	3.9 x 3.9
National Institute of Geophysics and Volcanology	INGV	Italy	SXG 2005	1.9 x 1.9
Meteorological Research Institute,Japan	NIES	Japan	MIROC3.2 (hires) MIROC3.2 (medres)	1.1 x 1.1 2.8 x 2.8
National Institute for Environmental Studies	MRI	Japan	MRI-CGCM2.3.2	2.8 x 2.8
Meteorological Research Institute of KMA	METRI	Korea	ECHO-G	3.9 x 3.9

Institution	Acronym	Country	Model	Resolution
Bjerknes Centre for Climate Research	BCCR	Norway	BCM2.0	1.9 x 1.9
Institute for Numerical Mathematics	INM	Russia	INM-CM3.0	4 x 5
			HadCM3	2.5 x 3.75
UK Met. Office	UKMO	UK	HadGEM1	1.3 x 1.9
Geophysical Fluid Dynamics Laboratory	GFDL	USA	GFDL-CM2.0,	2.0 x 2.5
			GFDL-CM2.1	2.0 x 2.5
			GISS-AOM	3 x 4
Goddard Institute for Space Studies	GISS	USA	GISS-E-H	4 x 5
			GISS-E-R	4 x 5
National Centre for Atmospheric Research	NCAR	USA	PCM	2.8 x 2.8
			CCSM3	1.4 x 1.4

Annex 10: Climate change perturbations for 2011—2030 and 2046-2065 based on the 1961-1990 baseline for the HADCM3 GCM.

HADCM3 GCM pertubations

[1]	month
[2]	relative change in monthly mean rainfall
[3]	relative change in duration of wet spell
[4]	relative change in duration of dry spell
[5]	absolute changes in monthly mean min temperature
[6]	absolute changes in monthly mean max temperature
[7]	relative changes in daily temperature variability
[8]	relative changes in mean monthly radiation

Scenario A1B 2011-2030

[1]	[2]	[3]	[4]	[5]	[6]	[7]	[8]
Jan	1	1	1	1.18	1.18	1	1
Feb	1.07	1	1	1.12	1.12	1	0.99
Mar	1.06	1	1	1.06	1.06	1	1
Apr	0.83	1	1	1.26	1.26	1	1.02
May	0.65	1	1	1.53	1.53	1	1.03
Jun	0.87	1	1	1.41	1.41	1	1.01
Jul	1.17	1	1	1.08	1.08	1	0.99
Aug	0.98	1	1	0.95	0.95	1	1
Sep	0.69	1	1	1.11	1.11	1	1.01
Oct	0.73	1	1	1.35	1.35	1	1.02
Nov	0.92	1	1	1.37	1.37	1	1
Dec	1	1	1	1.23	1.23	1	0.99

Scenario B1 2011-2030

[1]	[2]	[3]	[4]	[5]	[6]	[7]	[8]
Jan	1.08	1	1	1.15	1.15	1	0.99
Feb	1.05	1	1	1.01	1.01	1	0.99
Mar	1.04	1	1	0.99	0.99	1	0.99
Apr	0.93	1	1	1.13	1.13	1	1.01
May	0.81	1	1	1.23	1.23	1	1.02
Jun	0.81	1	1	1.16	1.16	1	1.02
Jul	0.91	1	1	0.97	0.97	1	1
Aug	0.88	1	1	0.94	0.94	1	1
Sep	0.64	1	1	1.24	1.24	1	1.03
Oct	0.6	1	1	1.59	1.59	1	1.05
Nov	0.85	1	1	1.63	1.63	1	1.03
Dec	1.05	1	1	1.4	1.4	1	1.01

Scenario A1B 2046-2065

[1]	[2]	[3]	[4]	[5]	[6]	[7]	[8]
Jan	1.08	1	1	2.37	2.37	1	0.98
Feb	1.17	1	1	2.13	2.13	1	0.97
Mar	1.09	1	1	2.14	2.14	1	0.98
Apr	0.77	1	1	2.65	2.65	1	1.02
May	0.49	1	1	3.1	3.1	1	1.05
Jun	0.68	1	1	2.98	2.98	1	1.02
Jul	1.2	1	1	2.5	2.5	1	0.98
Aug	1.31	1	1	2.19	2.19	1	0.97
Sep	0.99	1	1	2.3	2.3	1	1
Oct	0.81	1	1	2.57	2.57	1	1.02
Nov	0.92	1	1	2.63	2.63	1	1
Dec	1.03	1	1	2.53	2.53	1	0.99

Scenario B1 2046-2065

[1]	[2]	[3]	[4]	[5]	[6]	[7]	[8]
Jan	1.2	1	1	1.86	1.86	1	0.97
Feb	1.27	1	1	1.53	1.53	1	0.95
Mar	1.22	1	1	1.26	1.26	1	0.95
Apr	0.94	1	1	1.3	1.3	1	0.98
May	0.61	1	1	1.67	1.67	1	1.02
Jun	0.64	1	1	1.89	1.89	1	1.01
Jul	0.97	1	1	1.74	1.74	1	0.99
Aug	1.06	1	1	1.62	1.62	1	0.98
Sep	0.74	1	1	1.94	1.94	1	1.01
Oct	0.63	1	1	2.36	2.36	1	1.03
Nov	0.85	1	1	2.37	2.37	1	1.02
Dec	1.05	1	1	2.14	2.14	1	0.99

Annex 11: The Norera reservoir

The required storage of the reservoir is determined using the flow mass curve (Rippl diagram). A flow mass curve (Rippl Diagram) is a plot of the cumulative discharge volume against time. The equation for a mass curve at any time t is given by $V = \int_{to}^{t} Qdt$

Where
V = the accumulated volume at time t
to = the time at the beginning of the curve
Q = the discharge rate
t = the time
A mass curve was constructed by plotting the accumulative monthly flows against time. The slope of the mass curve at any time is a measure of the inflow rate at that time. Demand curves are straight lines having a slope equal to a – constant - demand rate. Demand lines drawn tangent to the high points of the mass curve represent rates of withdrawal from the reservoir.
The slope of the mass curve at any point represents

$$\frac{dv}{dt} = Q = \text{rate of flow at that instant}$$

Assuming the reservoir to be full whenever a demand line intersects the mass curve, the maximum departure between the demand line and the mass curve represents the reservoir capacity required to satisfy the demand. At the beginning of the dry period, the storage required is

$$\text{Storage (S)} - \text{max: of} \left(\sum \text{Volume demand} - \sum \text{Volume supply} \right)$$

The minimum storage volume required by a reservoir is the largest of Vsupply values over different dry periods. Assumptions in applying the mass curve method are that the reservoir is full at the beginning of the critical drawdowal period and, as the analysis utilizes historical streamflow data, it is implicit that future sequences of inflow will not contain a more severe drought than the historical sequence. The demand information (Table I0 was obtained from Hoffman (2007)

There is a water deficit in months of January- February. According to Hoffman, the abstraction of water from the Mara is limited in the drier months of July to September. The demand curve therefore serves only as a guide. Furthermore, the Amala is not the only tributary in the watershed. The Nyangores River contributes an equal or larger amount of water to the basin. The Nyangores river serves Bomet, the largest town in the basin. It is also the water supply for the Tenwek referral hospital. This study assumes that the water from the Nyangores will be used to provide for the increasing population in these two users and will not be used for irrigation.In addition, feasibility studies have been done only for the damming of the Amala. The long term data (Oct 1955-April 1995) for the Kapkimolwa gauging station (1LB02) (Table II) was used to construct the mass curve.The Amala River has been reported to experience both extreme low and high flows, and flood regulation is therefore a priority (Mango et al. 2012). For the mass curve generation the highest monthly demand (3647 ~3650cumec-day) was used. An assumption that this monthly demand will be the same throughout the year was made.

Table I. estimated monthly & annual consumptive water use (m³) in the MRB (modified from Hoffman, 2007)

Month	Total consumptive water user per month m3

January	2117814.9**
February	1765473.4
March	1521980.8
April	1413581.7**
May	1727056.3**
June	2247673.5
July	2713726.5*
August	2527675.2*
September	2302131.3*/**
October	2607258.7*
November	1531110.9
December	1336971**

*Assumes the annual migration is within the MRB for the four month period from July through Oct

**Abstractions not possible for Sept and limited in December

Table II. Accumulated monthly flow and demand for the Mara River at 1LB 02

Month	Observed flow, m3	Actual Demand, m3	Highest demand, m3	Acc. Flow m3	Actual Demand actual, m3	highest demand demand, m3
Jan	2167488	2117814	2713727	2167488.2	2117814	2713727
Feb	1569786	1765473	2713727	3737274.6	3883287	5427454
Mar	2766980	1521981	2713727	6504254.3	5405268	8141181
Apr	6808513	1413582	2713727	13312768	6818850	10854908
May	7847573	1727056	2713727	21160340	8545906	13568635
Jun	3788786	2247674	2713727	24949127	10793580	16282362
Jul	5183174	2713727	2713727	30132300	13507307	18996089
Aug	7512612	2527675	2713727	37644912	16034982	21709816
Sep	8490799	2302131	2713727	46135711	18337113	24423543
Oct	6488611	2607259	2713727	52624322	20944372	27137270
Nov	1891728	1531111	2713727	54516050	22475483	29850997
Dec	2064930	1336971	2713727	56580980	23812454	32564724

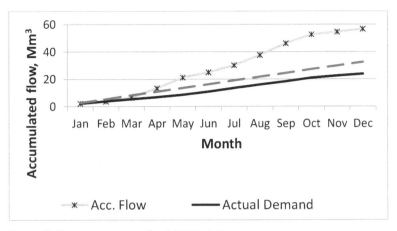

Figure I. Flow mass curve for 1L0B2 station

From the graph the required storage volume is 216Mm3. This volume was compared with the volume determined using the deficit/excess method given in table II below. In the deficit/excess method, three alternatives for determining the deficit were used. The actual difference scenario, a worst case scenario where the highest demand in the year is adopted as the monthly demand, and a case where the mean of the monthly demands is used. The worst case scenario gives 6 to 10 times the deficit as compared to other cases. The mass curve method gives a higher storage volume even in the worst case scenario. A storage reservoir with a volume of 400 Mm3 is adopted for the SWAT model irrigation from reservoir scenario analysis.

Table III. Reservoir storage determination from long term average monthly flows and the monthly demand for the Mara basin

Month	Days	Flow (cumecday) I	Demand (cumec-day) O	Actual	Deficit/Excess flow (Flow-Demand) worst- case	mean
Jan	31	3010	2847	164	-640	294
Feb.	28	2180	2627	-447	-1470	-536
Mar	31	3843	2046	1797	193	1127
Apr	30	9456	1963	7493	5806	6740
may	31	10899	2321	8578	7249	8183
Jun	30	5262	3122	2140	1612	2546
July	31	7199	3647	3551	3549	4483
Aug	31	10434	3397	7037	6784	7718
Sept	30	11793	3197	8595	8143	9077
Oct	31	9012	3504	5508	5362	6296
Nov	30	2627	2127	501	-1023	-89
Dec	31	2868	1797	1071	-782	152
Minimum storage required (Mm3)				39	338	54

15. List of publications

Peer reviewed publications

Kilonzo F., F. O. Masese, A. Van Griensven, W. Bauwens, J. Obando, and P. N.L. Lens, 2013. Spatial-temporal variability in water quality and macro-invertebrate assemblages in the upper Mara river basin, Kenya, Physics and Chemistry of the Earth, Parts A/B/C, Available online 26 October 2013.

A. van Griensven, P. Ndomba, S. Yalew, and **F. Kilonzo**, 2012. Critical review of the application of SWAT in the upper Nile Basin countries, Hydrol. Earth Syst. Sci., 16, 3371–3381.

Kilonzo F., A. Van Griensven, P. Lens, J. Obando, and W. Bauwens, 2012. Improving SWAT model performance in poorly gauged catchments, Centre International Capacity Development (CICD) Series (ISSN 1868-8578).

Kilonzo F., A. Van Griensven, and W. Bauwens, 2012. Distributed Validation of Hydrological Model Using Field and Remotely Sensed Data. In Seppelt et al., (Eds.): Managing Resources of a Limited Planet: Pathways and Visions under Uncertainty, Sixth Biennial Meeting, Leipzig, Germany, International Environmental Modelling and Software Society (iEMSs).

Conference proceedings

Kilonzo F., and J. Obando, Food production systems and the farmers' adaptive capacity to Climate Change in the Upper Mara River Basin, Kenya, 3rd Lake Victoria Basin Scientific conference, 2012, Entebbe, Uganda. book of Abstracts pg 33.

Kilonzo F., F. O. Masese, A. Van Griensven, W. Bauwens, J. Obando, M. McClain, and P. Lens, 2012. Spatio-temporal variation in water quality of the upper Mara River basin, WaterNet/ Symposium, Johannesburg, South Africa book of abstracts.

Kilonzo F., A. Van Griensven, P. Lens, J. Obando, and W. Bauwens, 2011, Selecting a Potential Evapotranspiration method (PET) in the absence of Essential climatic input data, SWAT International Conference Toledo, Spain, proceedings, pgs 438-446.

Kilonzo F., A. Van Griensven, P. Lens, J. Obando, and W. Bauwens, 2010. Improving model predictive capacity in poorly gauged catchments, LAKE AND CATCHMENT RESEARCH SYMPOSIUM (LARS2011) conference book of Abstracts (LARS2011-pg 47).

LIST OF PUBLICATIONS IN THE SERIES "VUB-HYDROLOGY"

EDITED BY THE DEPARTMENT OF HYDROLOGY AND HYDRAULIC ENGINEERING

c   Dienst Uitgaven VUB, Wettelijk Depot 1885

fax +32-2-6293022, e-mail: hydr@vub.ac.be,http://www.vub.ac.be/hydr

About the author

Kilonzo Fidelis Ndambuki (°24 Jul 1972, Kenya) obtained his Bachelor diploma in Agricultural Engineering at the Jomo Kenyatta University of Agriculture and Technology, Nairobi, Kenya in 1996. In 2003, he obtained a Master diploma in Water Resources Management from the Lueneburg University in Germany. Finally, he got a Master diploma in Environmental Sanitation from the University of Ghent (Belgium) in 2005. He worked on his Phd from 2009 to 2013, with the support of the University Development Cooperation (VLIR-UOS) in Belgium and from Nuffic in the Netherlands. From 2008 onwards he was also part time lecturer at the Faculty of Engineering of the Jomo Kenyatta University of Agriculture and Technology and tutorial fellow at the School of Engineering and Technology at Kenyatta university.

T - #0389 - 101024 - C306 - 244/170/17 - PB - 9781138026384 - Gloss Lamination